WHEN WOOL WAS KING

WHEN WOOL WAS KING

The inside story of Australia's wool industry

Alec Morrison AM

R.M. Williams
PUBLISHING

Published in Australia in 2009 by
R.M.Williams Publishing Pty Ltd
ACN 068 461 263
Level 11, 52 Alfred Street
Milsons Point NSW 2061

www.outbackmag.com.au

First published 2009

National Library of Australia Cataloguing-in-Publication entry

Author:	Morrison, Alec, 1935–2004.
Title:	When wool was king: the inside story of Australia's wool industry / Alec Morrison.
Edition:	1st ed.
ISBN:	9780980397031 (pbk.)
Notes:	Includes index. Bibliography.
Subjects:	Wool industry – Australia – History. Wool – Australia – History.
Dewey Number:	338.1763140994

Cover design by Jeremy Nicholson, Bland Design
Cover image: Courtesy of the Morrison family
Designed and typeset by Kirby Jones
Printed and bound in Australia by Griffin Press

Dedicated to the loving memory of Alec Morrison
(1935–2004)

FOREWORD

The wool industry, and in particular Merino sheep, are a crucial financial and cultural factor in the history of Australia. *When Wool was King* is an insightful, personal record of the role of wool in the formation of a country so dependent, in both the bush and the city, on this amazing fibre and the business of growing and selling it. Alec Morrison's own career saw him progress from his early life as a jackaroo at the start of the 1950s to eventually become one of wool's leading lights. His experience and understanding of all aspects of the wool game make him a most credible person to write a book devoted to this epic era – from the 1920s to the turn of the 21st century – of a pivotal Australian industry.

Alec was a romantic and thoughtful observer and recorder of all that went on around him during a period of great growth and great change. He was influenced by a long and impressive family involvement in the bush and sheep and brought a keen intelligence and genuine empathy to writing this book. I had the honour of working closely with him in presiding over F.S.Falkiner & Sons, where Alec showed enormous flair in his leadership of this famous sheep and wool operation.

Moreover, his involvement and commitment extended beyond the boundaries of F.S.Falkiner's holdings. Alec was a founding director of the Australian Wool Research and Promotion Organisation and also served as chairman, with his wit and admirable capacity for hard work combining to great effect in this very challenging and sensitive wool industry body.

When Wool was King is virtually the story, professionally at least, of Alec Morrison's very special life, and we should all embrace it.

Ken Cowley AO
June 2009

CONTENTS

PREFACE

For most of our history the only thing we did better than anyone else in the world was to grow wool – sheep stations and station life helped define our national character and how we saw ourselves. That lifestyle has gone, as have most of the big stations. This book is an attempt to describe to later generations what station life was like for almost a century.

I am a third-generation sheep station man. My grandfather owned a station in the Cobar district of New South Wales until he went broke in the late 19th century due to a combination of drought, incompetence and bank closures. My father was a sheep station manager for most of his life. I worked on and knocked about sheep stations for the best part of 50 years.

A unique social structure, management system and culture had started to form on the stations as early as the 1850s, in the days of the squatters. Because of their size and isolation, the big stations became self-contained, self-sufficient communities of their own, but nationwide their people shared a common lifestyle, common attitudes, and a common dedication to work. Station people from around Australia could recognise one another instantly – there was something distinctive about the style of a station man.

The old structure and culture began to crumble in the 1960s and 1970s under pressure from relentless social and economic forces. The wool crisis of the 1990s swept up what was left, so by the end of the 20th century the old ways had virtually disappeared.

When I was a young man, and the Australian population was smaller, nearly everyone living in the coastal cities had some bush connection. One of the family may have owned a farm, or they knew someone who had been a stockman, a shearer or a roustabout, or a rabbiter, an overseer or a bookkeeper, or a jackaroo or a manager. That situation no longer exists. At the start of the 21st century Australia has inevitably grown away from its agricultural roots and its pioneering icon industries such as wool growing. A city-bush connection does still exist, but in a different way. Better roads, motel accommodation and ownership of four-wheel-drive vehicles ensure more of the population knows what the outback looks like, but fewer understand what it was like to work and live there when wool was king.

Alec Morrison AM
Melbourne, 2004

ACKNOWLEDGEMENTS

Alec began this book on the social history of the Australian wool industry after his retirement in 2000. He went on to write several other books before returning to complete it prior to his death in July 2004. Alec's book charts the history of the wool industry from the 1920s to 2000. This volume has an additional chapter researched and expertly written by freelance rural journalist and author Amanda Burdon, telling the story to the present. To keep the voice of the text, Amanda contacted many of Alec's associates for their insight into how Alec would have viewed the recent changes to the wool industry.

Many people who share Alec's passion for this industry have assisted by contributing their stories and experiences, as well as giving practical help to see this history come to fruition. Alec valued all of these contributions. Thank you to the following people who generously gave Alec their time for interviews: Sir William Allen, of Brisbane; Hugh Beggs, of Glenthompson, Victoria; Malcolm Brown, of Jerramungup, Western Australia; the late Doug Carty-Salmon, of Melbourne; Roly Desaily, of Hay, NSW; Michael Elmes, of Narrandera, NSW; Neil Graham, of Melbourne; the late Mark Grigson, of Sydney; Adrian Kloeden, of Melbourne; Bill McLaughlin, of Molong, NSW; the late Bill McPherson, of Wagga Wagga, NSW; Bill McPhillamy, of Forbes, NSW; Jon Sendall, of Noosa, Qld; David Ward, of Melbourne; and Jim Young, of Melbourne.

Thanks also to Alec's sister, Monica Rutledge, Louise Sweetland and the late Tim Hewat for their enthusiastic encouragement and incisive comments at the beginning of Alec's writing; and to Michael Elmes, Laurie Piper, Tim Jones and Michael Hanrahan for their contributions after reading the completed manuscript. Thank you, too, to the other people, too numerous to name here, involved over the life of this work.

One person whose friendship and involvement in Alec's life and this work I would like to particularly acknowledge and thank is Ken Cowley. Ken was Alec's boss at F.S.Falkiner for more than 20 of the 47 years of Alec's working life, encouraging Alec's pursuit and love of working in the wool industry. Following Alec's death, Ken, through his company R.M.Williams, honoured Alec's lifetime contribution to developing and preserving the outback with the Outback Heritage Award in 2004. More recently it is thanks to Ken that R.M.Williams took up the publication of this book. I would like to thank the team at R.M.Williams Publishing, especially Therese Hall, whose assistance has been invaluable in negotiating the complexities of bringing this work to print.

For permission to publish many images for this book I would like to thank for their co-operation Colin Bell and Robyn Grose, F.S.Falkiner & Sons; Denise Thomas, Deniliquin Peppin Heritage Museum; Bill Mulham, Deniliquin Historical Society; George Falkiner, *Haddon Rig*, Warren; the late Ross Field, T.A.Field Estates, Pty Ltd; and Graeme Black, *Uardry*, Hay.

Finally, thanks to the special people in my life Robert, James and Lisa; my sisters Jet, Layla and Dorla; and my wonderful friends for their support and advice.

Em Morrison
Melbourne, June 2009

The sheep was one of the first animals created,
Possibly the very first,
Since it could amply supply for mankind all that was most necessary –
Meat, milk and clothing.

From an old English proverb

CHAPTER ONE

JACKAROOS

If you want a situation, I'll just tell you the plan
To get on to a station, I am just your very man.
Pack up the old portmanteau, and label it Paroo,
With a name aristocratic – Jimmy Sago, Jackaroo.

ANON, FROM AB PATERSON'S *OLD BUSH SONGS*

Out of all this wild confusion,
After meditation deep,
I have come to the conclusion
That the road to hell is sheep.

ANON, *THE JACKAROO'S LAMENT*

One steamy January morning in 1953, just turned 18 and straight out of school, I boarded a rackety piston-engine aircraft at Mascot in Sydney, bound for Brisbane and thence Blackall in central-western Queensland. It was my first trip in an aeroplane, and I was to vomit solidly for the whole journey. Saying goodbye to my parents was a trial, but we were all used to it. Bush children of my generation, and my father's, had been packed off to boarding school at 10 years of age or younger, and our early life was a constant round of separations. Who of us will forget the awkward conversations on bare and scorching railway platforms, dreading yet looking forward to the moment of departure, suppressed tears all round, manly handshakes. In those days, long-distance telephone calls were an expensive luxury and the cost of travel was prohibitive, so we all knew we would not be speaking to one another again for many months, probably a year.

Like my father before me, and thousands of other young men over the years, I was heading off to a back-country sheep station as a jackaroo. Like all the others, I was ambitious and firmly convinced that the career I was embarking on had some significance in the wider scheme of things. After all, we were starting out in what was then Australia's first and greatest industry – wool growing. We carried reading matter with us; usually a book of Banjo Paterson's poems, the complete works of Shakespeare, and 'The Jackaroo's Bible' – *Sheep, Farm and Station Management*, a book of practical advice compiled from the columns of the monthly journal *Pastoral Review*. Its opinion about our worth was not encouraging: 'Young men going to stations to gain experience must realise that, for a commencement at least, they are of little value, and only tolerated by the managers. An "Experience Man" is not as a rule worth very much remuneration.' The book was right – most of us were incredibly green and innocent, and so were of little use until our second year. The author also issued a stern injunction: 'Don't drink spirituous liquor until you are 30, and all you drink after that will not hurt you.' Most of us were still wearing school clothes, all ankles and wrists and bursting at the seams, though the more affluent may have had an immaculate new sports coat tailored by Chorleys of George Street in Sydney as a going-away present. Most of us wore a spanking new pair of elastic-sided riding boots and a broad-brimmed felt hat. We were issued a quart pot and saddlebag from the station store, with the cost deducted from our first month's wages. Our second month would see us sending off for mail-order riding trousers and khaki shirts from R.M.Williams or taking advantage of a hawker's visit.

We had cleared a number of hurdles in finding a suitable billet with career prospects. Religion was an issue in those times. Though it was never acknowledged publicly, some stations took only Protestant boys, others only Catholics. To come from a 'good' middle to upper middle class family was important, as was good schooling. Attendance at one of the major private schools was almost essential for a position on the better stations; at one stage only old boys of The King's School could get a billet at the Body family's *Bundemar*. During my time at Blackall only two of the many dozens of jackaroos in the district had attended state schools, and in our patronising way we felt rather embarrassed for them. Academic results were considered of minor importance, but being in the top sporting teams or holding the rank of prefect counted; leadership qualities were seen as paramount, but in the physical rather than in the intellectual sense.

The foundations of the jackaroo system had been laid around the middle of the 19th century. Wealthy families with a mind to invest in the pastoral industry, or English families seeking a place for wayward younger sons, placed their boys with the leading station managers of the day to learn the business. Then known as 'Colonial Experience Men', they at first had to pay for the privilege. In later years nominal wages were paid, but, like subalterns in the British army, jackaroos were expected to supplement their meagre pay from private means. Some were well able to do so – there was a young man at *Boonoke* in my time who drove a Bentley. When my father started at *Brookong*, near Lockhart in New South Wales, in 1913, he received five shillings a week and was expected to provide his own saddle and gear.

Like navy midshipmen in earlier times and budding hotel executives in modern times, a jackaroo was expected to learn at least the rudiments of every job carried out on a property. In addition to stockmanship he was expected to learn to shear a sheep, kill and dress meat, repair fences, swing a new gate, strip down engines, repair windmills, pull a bore, drive a grader, carry out electrical repairs, maintain a telephone system, do emergency plumbing, and understand the basics of double-entry bookkeeping. It was thought necessary he should know all these things so that when he became an overseer and eventually a manager he would be able to supervise standards of workmanship and know how long any particular job should take.

Jackaroos, managers and overseers were seen as part of a large 'officer class', and families fully expected their jackaroo sons not only to be taught the profession of station management, but that the manager should see to their social, moral and ethical development. If a boy was to commence a sexual relationship with the wife or daughter of a neighbour, and of course it happened, it would be seen as reprehensible behaviour but not of sufficient gravity as to warrant dismissal from the officer class. But if a liaison developed with a domestic servant then it crossed class boundaries, and that person bore the stain for the rest of his career; his prospects would most certainly have been limited. The manager of one large and famous property in central-western Queensland during the 1950s had, some 20 years earlier, married the station cook. They were still being excluded from many district social events and were denied membership of the local club.

Social niceties and class demarcations were strictly adhered to on stations prior to the 1960s. Each new jackaroo was given a lecture by the overseer or manager about the need to maintain cordial relationships with 'the men' – meaning station hands and boundary riders – but on no account were they to mix after working hours or at social events.

We lived in 'barracks', the men in 'huts'; we ate in a 'dining room', they in 'the mess'; we ate off china plates and drank from china cups, they from enamel. The overseer or bookkeeper acted as head of the household, sat at the head of the dining table and was expected to ensure that the jackaroos behaved themselves. Linen, laundry, ironing and meals were all provided by the station, and by the time I started in 1953 wages had progressed somewhat to the princely sum of five pounds a week, increasing to £10 after four years' experience. Life in barracks was an easy transition from boarding school for boys used to robust male company, so much so that many young men who left school and went on to work in a large pastoral company for most of their careers became more or less institutionalised, to the extent that they were never happy working outside the certainties and routines of company life.

Our days were regulated by a regime of bells – 'getting up' at 6.30 am, breakfast at 7 am, lunch at noon, 'dressing' at 6.30 pm, dinner at 7 pm. Jackaroos usually took at least one meal a week with the manager and his family, often at weekends. There a degree of formality was observed, with a housemaid in a black dress and frilly cap, or a houseboy in a mess jacket serving at table. A coat and tie were obligatory, or evening dress in pre-war times. Some of the English companies stipulated that jackaroos should dine each night with the manager.

Our social life was severely constrained by the constant demands of work. It was a monastic existence, with little opportunity to develop a sex life. Trips to town were few because no-one owned a car, and in any event there was too much competition for the few single girls from the droves of single men in from the district stations. A former jackaroo recollected his time at *Wellshot*, near Longreach in Queensland, during the 1930s: 'We didn't get to town very often. Town visits were not encouraged and in any case we were too busy. I remember calculating that in one year we worked 26 Sundays.'[1] Another recalled some trivial domestic incident while he was at *Westlands*, also at Longreach, in the 1920s that illustrated the closed nature of station life: 'You could say these small incidents were of no consequence, but we had rare visits off the place, even to Longreach.

There was no wireless or television, little spare time to read, and our thoughts did not go far afield; we lived in a small world of our own, and they gave us something to talk and laugh about. It was better than growling about the heat, dust and flies.'[2]

When cooped up in barracks during wet weather, life could be trying for those who had not armed themselves with some timeless reading matter. One jackaroo at *Warbreccan*, 240 kilometres south-west of Longreach, in the 1930s remembered: 'Very foolishly we read every paper the same night the mail arrives, so "in the wet" there may be nothing else for several weeks, except the labels on the sauce bottles and jam tins.'[3] I had found myself so desperate for reading matter that I was forced to go through the drawers of furniture to tear out and devour the old newspapers used as lining.

A fresh face at the dining table was always most welcome and provided new topics of conversation. At *Terrick Terrick* in the 1950s we once had a visit from a Christian missionary of the more extreme sort, who stayed overnight and harangued us endlessly about his particular beliefs. With

HM Nash and Alec (at right): young jackaroos on *Terrick Terrick*, Blackall, Qld, in 1953.

little else to do, we argued the toss and contested his every point. It was all good fun, and at the end of the night we each bought a book from him. Fifty years later, mine still sits on the shelf unread. It cost me a week's wages. When he finally left, he stood at the door and cursed us roundly in a most unchristian manner, and what's more told the people at the next station we were a godless lot who most certainly would fry in hell when 'The Day' came.

What little society there was for jackaroos usually revolved around the station's cricket team, amateur horse races or the occasional tennis party, which often entailed considerable travel. One young man recalled a trip from *Warbreccan* in 1932: 'We rose by 5.30 am on Saturday, rolled our swags and set off after breakfast at six – Jim in the truck with the overseer, George, Max and I with the Dovetons in the Dodge. After four nips at the Jundah pub, we had morning tea at *Moyen* and reached *Galway* for dinner. The house is up on large stilts, with a large cool living area underneath, where food and drink were set out. Good tennis during the afternoon, and dancing at night, between plenty of grogs. A fierce burning gale spoilt tennis on Sunday morning, so we spent the time playing the gramophone and looking at magazines. We left at 4.20 pm, and after a few last nips at Jundah were home for dinner at 8.'[4]

During my time at *Terrick Terrick* no jackaroo owned a car, and we got into Blackall probably three or four times a year: for the annual show, where we held sheep; to a race meeting; and perhaps a dance or two. If you were lucky there was a trip to Longreach, or even the metropolis of Charleville, with the team of show sheep. Some unkind soul once said the *Terrick Terrick* jackaroos were drunk every time she saw them, forgetting that this was only once a year. Of course, all this was to change within 10 to 15 years, with increased ownership of cars and improved roads.

Wherever you travelled in the bush in those times the music for dances seemed to be provided by a similar group: a stout woman of a certain age, wearing a black sequinned dress and pounding an upright piano with plump, bejewelled fingers; an elderly saxophonist wearing an ill-fitting hairpiece and a white tuxedo; and a bored drummer. Their repertoire was limited to about five numbers, including a quickstep, slow foxtrot and jazz waltz, all of 1930s vintage. But everyone had a great time, and no-one saw anything inadequate or mean in what surrounded us.

Though there were a few 'soft camps', work for jackaroos was mostly hard and unremitting. A former employee at *Wellshot* recalled: 'Before taking the position, I asked one old hand how many days a week they worked there. He said "14", and I really do not think he made any error. We used to get up at 4.30 am and after breakfast went out and sat on the rails until it was light enough to see our horses.'[5] Most jackaroos, at least those on a career path and without a family property to return to, took their work very seriously. There was fierce competition for promotion and preferment: 'There was always the spur of competition. We used to match ourselves against one another all the time. If a sheep had to be killed there was a standard time, and you were always trying to better that time. At the horse yards if you got a young horse and you got thrown, you got back on because there were people watching. You kept getting back on no matter how many times you got pelted because it was a matter of pride.' Despite the poor pay and long hours, ambition ensured there were no industrial protests, except for one famous incident at the Falkiner's *Haddon Rig* in the 1950s. In the manager's absence the overseer fired one jackaroo. The other jackaroos believed this was done unjustly, and they put in an 'if he goes we all go' ultimatum. On his return the manager called their bluff, as he was bound to do, so all rolled their swags and headed for the local town of Warren. The manager made a proviso that anyone who was prepared to apologise would be reinstated. All but one maintained solidarity. That individual argued, with some justice, that he was a career jackaroo without a family property to return to, so could not afford a black mark so early in his working life. Many of the people involved in the incident went on to become prominent figures in the pastoral industry, but they never forgave the one who stayed on, and for more than 30 years none would recognise nor speak with him.

The most sought-after billets for a young man of ambition were those in the great pastoral companies that owned strings of properties – Australian Estates, New Zealand and Australian Land Co., Scottish Australian, Squatting Investments, Australian Pastoral Co., Goldsbrough Mort, and so on. They offered a career path through to manager and even beyond. My generation were inspired by people such as Duncan Mathieson, who had progressed steadily from jackaroo to overseer to manager to general manager of the Scottish Australian Company. Promotion through the ranks for the keen and competent in the companies was sure but slow – from jackaroo to overseer often took eight to 10 years, then another 10 to reach manager status. A 'company man' seldom had the opportunity to marry until he was nearly 40.

The companies took great care in the training of their jackaroos. Appointments were usually made at a senior level in head office, and boys were transferred between stations periodically so they gained experience in different types of country, climates, and a variety of enterprises. Few managers had the authority to sack a jackaroo without reference to head office. In my time with Australian Estates, each of us was required to keep a work diary, attend monthly briefings with the manager, undertake a station bookkeeping course, and attend schools on things such as sheep genetics and nutrition. Managers reported quarterly to head office regarding each boy's progress in the various practical skills, such as counting and drafting sheep, and on the development of their character. In 1933 *Barratta* manager CJC Cameron was having some problems with a jackaroo, as he reported to the A.M.L.&F. Co. in Sydney: 'He is a willing and quick worker, and at all times does his best in this respect, and is a boy with a lot in him, but requires very firm handling to properly bring out in him what is essential in a man. He is to a large extent too egotistical, and strange to say easier led by those against their employer's interest, than he is by the ones interested and responsible. I have however recently explained all to him, as a father to a son, which I am hopeful will have the desired effect.'[6] It must have worked because the subject of that report went on to become one of Australia's leading businessmen. Probably beyond redemption was the jackaroo reported on by one of the Australian Estates' more courageous managers: 'He is bone lazy, untidy, unshaven, unclean, a domestic nightmare, won't befriend anyone, hard on horses, impossible with machinery, and difficult to trust with a second-hand shovel. I feel he is only fit, and I thoroughly recommend him, for a position of responsibility in head office.'[7]

Initial promotion to senior jackaroo or overseer was based on a combination of seniority and physical input – those who worked hardest and fastest got on first. This frequently threw up problems as men moved from playing a purely physical role to taking on thinking and planning activities. I have seen young men – superb craftsmen out in the paddocks – grapple with using a telephone, making up livestock inventories, and working with suppliers. Promotion to manager soon revealed any intellectual shortcomings, particularly at critical times such as dealing with drought conditions. Promotion from the ranks was rare prior to the 1970s, but there were outstanding examples of how a more liberal attitude might have worked. Alf Walker recalled his time in the Longreach area during the 1930s: 'The fact that Mr Murray agreed [for Walker to become overseer] was typical of his broad-mindedness. It was a big step up for me, because I had never been a jackaroo, always just a station hand, and I didn't have the social background that overseers had to have. There were thousands of good men who never got a chance because they didn't have the background, so I was very lucky. Once I had the job I knew I had to hang onto it.'[8] The Falkiner family made an inspired decision when they appointed former boundary rider Billy McKees to manage their Riverina holding *Boonoke* during the Second World War; he proved to be an outstanding stockman with sound judgement and boundless energy. His habitual dress consisted of lace-up boots, heavy twill work trousers pulled up to breast height by police and firemen's braces, a thick blue cotton shirt with short sleeves, and a great floppy felt hat like a mushroom. His only concession on public occasions was to wear a white shirt. McKees had received little education, but was naturally clever, and possessed a great turn of phrase. I once heard him describe the sense of isolation he felt when thrown from a young horse out in the back paddock of *Boonoke*. 'There I was – all alone – out on the plain – by myself – no bugger with me!'

The freeing up of social attitudes that attended the 1960s, and the wool recession of the late 1960s and early 1970s, changed the jackaroo system forever. The reputation of the wool industry had been eroded by bad publicity, and stations that once had long waiting lists for jackaroos were finding it hard to find applicants. The big pastoral companies were being broken up. A different portfolio of management skills was now required to cope with poorer terms of trade in the pastoral industry, and practical training became less relevant. Tertiary education was becoming more widely available, as were other career opportunities at higher rates of pay. A more egalitarian and less class-conscious society meant that careers no longer depended upon maintaining the status of gentlemen, and promotion became based on performance and merit. Increased mobility in both a physical and social sense meant that stations no longer operated as closed communities. The future of management training in the pastoral industries lies with a mixture of tertiary education, delivered via the new technologies, and practical experience. The station manager of the future will need technical skills, the ability to deal with masses of information, and a sound knowledge of the general principles of management.

Despite its class-based nature, the old jackaroo system had many good features in its time. By placing a lot of pressure on young men to perform, it taught them a strong work ethic, high standards of workmanship, self-discipline, the value of teamwork, the need for persistence, and moral and ethical values. And it was a great life. People developed lifelong attachments to their first billet. During the nearly 30 years I spent running the F.S.Falkiner sheep empire in the Riverina of New South Wales there was a constant procession of middle-aged and elderly men from all walks of life, many of them highly successful, making a sentimental pilgrimage to the stations where their careers began. They had never forgotten what those early experiences had done for them in building their character and their values.

When I arrived at *Terrick Terrick* that night at the start of 1953, in teeming monsoonal rain, and smelt the strange aroma of wet gidyea trees, I felt a sense of eager anticipation. I little knew that it was the start of a 50-year journey, a journey that would contain many peaks and troughs, a journey during which I was to meet many great characters, a good few rogues, and a cast of passionate and committed people. During those 50 years I was to see the station life that I knew, and my father knew, change forever, and eventually disappear.

CHAPTER TWO

FAR-FLUNG FIEFDOMS

The optimist proclaims that we live in the best of all possible worlds; and the pessimist fears this is true.

JAMES BRANCH CABELL (1879–1958)

The locals called it *Big Brookong*. The station lay between the towns of Lockhart and Urana in New South Wales. Reduced from its original 85,800 hectares, in 1920 it was still a hefty 26,700 hectares carrying 40,000 sheep. Within two years it would be gone, split into small blocks and sold off by the owners, the Australian Estates and Mortgage Company, in a deal with the government to protect their string of stations from further resumptions for closer settlement. *Big Brookong* was a famous property that today exists only in folklore and the memories of the few old men left who lived there. Physically all that is left is some tumbled and weatherworn timberwork, a sad relic of the mighty woolshed where once nearly 10,000 sheep were shorn in a day.

In the autumn of 1920 the newly appointed overseer, holding two handsome kelpie dogs, had his photograph taken. Typical of thousands of young bushmen, he had enlisted in the first Australian Imperial Forces (AIF), sailed to Egypt in 1915, gone on to Gallipoli, then France, returning in early 1919 physically and mentally damaged to spend six months doing little other than staring apathetically out the window of his mother's house in Melbourne. His symptoms were those of mild shellshock – 'a condition of alternate moods of apathy and high excitement, with very quick reaction to sudden emergencies but no capacity for concentrated thinking'.[1] He was 24, young for an overseer in those days, but having survived some tough situations, he looked older. Tucked in his wallet was a letter on buff military paper, which he was to carry for the remaining 37 years of his life.

No. 615, Sergeant JA Morrison, C Company, 23rd Battalion, 2nd Division, AIF.

This NCO acted as Platoon Sergeant during the operation near Beaurevoir, east of Peronne, on the night of 3/4th October, 1918. He displayed extraordinary coolness and courage under heavy machine gun and artillery fire. By his example, he was of invaluable assistance to his platoon commander in getting the men safely through the enemy barrage and on to their objective. On reaching the objective he made a valuable forward reconnaissance in daylight. His tireless energy and excellent conduct were of the highest standard and acted as a great incentive to his men. His Majesty the King has graciously approved the award of the Military Medal for bravery in the field.

(Signed) JC Robertson, Brigadier-General, Commanding Second Australian Division.

The overseer in the photograph is my father. He is clearly dressed for work, yet a sports coat is draped across a nearby fence, and he wears a tie. The station life he stepped back into had changed little since the 1890s. Indeed, it was almost indistinguishable from the one I knew in the 1950s. It was formally structured with strict codes of rank, dress and behaviour. Managers held powers akin to those of a ship's captain, with authority over the lives of all on the station. The manager of *Brookong* at that time was DCF ('Alphabet') Brown, who had protruding eyes and was not very popular among his staff. The jackaroos sang a version of the well-known song from 'Maid of the Mountains':

In days of yore I do recall
The dear old Brookong Creek
Where Duncan Brown paid us a crown
For all we did each week
If on this station I must stay
His eyes will fill me with dismay
And Jackaroos must take to booze
'Cos that's the only way.

Leisure time was limited to Sundays, and even that was provided no livestock needed attention. Employees had no access to private transport. There was a single party-line telephone in the office, and all were dependent on the station store for rations. Tactical management of the station was the manager's responsibility. He reported monthly to head office and was subject to quarterly visits from a pastoral inspector. Implementation of the daily work program, allocation of tasks, and organisation of staff fell to the overseer. Some stations might have a head stockman as next in rank, whose job it was to organise the mustering camps. Managers were addressed as 'Sir', overseers as 'Mister', and each wore a tie, often a coat or even a three-piece suit, as a badge of office which indicated they had passed beyond the necessity for physical work, even though at times they might work very hard indeed. After a long day out in the paddocks the manager and overseer met in the office after dinner to review the day, ring the outstations, plan the following day, write up work diaries and tend to administration. This might have taken them through to 9.30 or later.

The clear social distinction between officers and men that persisted through to the 1960s was firmly in place. A degree of social formality and restraint was exercised in those days that the more demonstrative 'baby boomer' and later generations would find strange – a 'real man' scorned the expression of emotion, and boastful or excessive behaviour was frowned upon. These formalities may appear ludicrous to those who live in more egalitarian times, but they had some basis. Few who live in urban communities and who have not served in the military fully understand the stresses of 'living over the shop' in a demanding physical environment, isolated and surrounded by the same people day in and day out. Some degree of structure and restraint is needed to keep the social fabric from tearing open. Even then things would sometimes crack. When I was manager of *Boonoke North* in the late 1950s we ate at a communal table. Everyone was preoccupied with work, and meals were usually rushed through in stony silence. Among our company was the station bookkeeper, a sweet middle-aged woman with rather prim and proper manners. The problem was that she had a set of ill-fitting false teeth that clicked audibly and persistently. Soon everyone was waiting in dread for the next click. After some weeks this became all too much for one of the jackaroos, who suddenly sprang up, flung down his knife and fork and shouted, 'Clickety-clack, clickety-clack, clickety-clack, I can't stand it

Alec's father, Jim Morrison: overseer on *Brookong*, Urana, NSW, in 1920.

you bloody old bitch', and stormed out of the room, thereafter taking all his meals on the verandah. At *Terrick Terrick* I developed such an antipathy to another jackaroo that I couldn't bear to look at him, and sat side-on to the table to avoid doing so.

A bush culture had been forming since the days of early settlement. Most station workers in the 1920s were single males who avoided polite society, scorned domestic comforts as being 'soft', and rejoiced in hard physical toil, which they saw as character forming. They actually enjoyed hardship, and above all other things admired self-sufficiency, physical toughness, and loyalty to 'mates'. The well-established stations did, however, attend to the comforts of the main homestead. A journal of the 1920s noted: 'Mr Otway Falkiner resides at the new homestead at *Boonoke North*, situated on the Colombo Creek, hence the water supply is good, evidence of which is borne by the wealth of green in the garden and grounds. In the large smoking room during the evening, after a busy day with buyers and non-buyers inspecting sheep, Mr John Carse, the Manager, hears reports of all the day's work per telephone.'[2] At *Boonoke*, a Falkiner family holding 140 kilometres to the south, the brick homestead boasted seven double bedrooms, which were always filled by visitors. Though there was no running water to the bedrooms, ewers of hot water were carried to each room morning and evening. Indoors there was one bathroom and one WC, which must have been busy, but relief was available for the urgent in the form of an earth privy in the garden. Legend has it that it was in the kitchen of this old building 30 years earlier that the teenage Otway Falkiner, later to become family patriarch, had lost his virginity to the Irish cook. He had been sent by his parents to help with the washing up, and was teasing and pinching her. Without warning she shot out one beefy arm to squeeze out the wick of the slush lamp, and with the other flung the unprotesting Otway onto the kitchen table and fell upon him. In view of his subsequent career, later generations of the family wanted the table to be fitted with a brass plaque with appropriate inscription.

Fred Knight, whose mother was a Falkiner, wrote of *Boonoke* in his memoirs: 'It might have appeared that the upkeep of the household was something like that of a good hotel, and so it was. This was not because Grandfather [Frank Sadlier Falkiner] was ostentatious – that would have been far from the truth. The reason was entertainment of ram buyers. The buyers of those days had long and sometimes uncomfortable journeys from Melbourne or Sydney, so they stayed for a week or more.' Hospitality seemed to work. One ram buyer told Fred Knight in the 1930s: 'It didn't matter to me what big stud I bought rams from. They all came from the same family [the Peppin strain of Merinos founded in 1861], so I bought where I was made most comfortable.' Not all visitors were of the gentleman class. There was an active trade in stud sheep with South Africa at the time, and the visiting Boer farmers smoked a particularly evil-smelling type of tobacco, and were inclined to spit on the floor. Many of the leading sheep studs built special smoking rooms so the household could be free of the noxious fumes from their pipes. Knight tells of one ram buyer of Scottish birth who had driven 50 kilometres in scalding heat and a dust storm. When asked if he would like a wash, he replied, 'Thank you, no, I had one before I left hame'. Later he was given a glass of champagne, but he was used to stronger stuff and, taking a mouthful, spat it out on the carpet in horror, snarling 'Pah! Lemonade!'

Dinner in the big homesteads was usually formal. At *Lake Midgeon*, near Narrandera: 'Dinner was a great ceremony, preparations for this commenced with a dressing gong which was rung at 6.30, everyone would return to the drawing room, changed and ready to file into the dining room when the second gong went at seven. A visitor once told me that one of her most vivid memories of *Midgeon* was that if you happened to have a drink in your hand you flung it down your throat and sprang to your feet to go into dinner before the gong had barely stopped booming. The staff must not be kept waiting – father's edict. There were always three courses: soup, meat with potatoes and two vegetables, then a pudding or perhaps a savoury.'[3]

These establishments required a large domestic staff. At *Boonoke* the inside staff consisted of a housekeeper, a male cook (usually Chinese), a laundress, four housemaids and a gardener. The Austins at *Lake Midgeon* were even better set up, with nanny, assistant nanny, ladies' maid, governess, chauffeur, parlour maid, cook and under-cook, kitchen maid, housemaid, laundress, barracks housekeeper, two general gardeners, Chinese vegetable gardener and a butcher.

One cannot but feel for the lot of domestic servants on the stations in those days – poorly paid, often overworked, few prospects, living in spartan quarters, no access to society other than their own, no entertainment, and no distractions. It was no wonder that things did not always go smoothly on the domestic front, and the two main Sydney domestic employment agencies, Bentleys and Copes, did a brisk trade interviewing, making wage advances, and sending people out on the trains. Life for the Chinese vegetable gardeners was especially difficult. They usually operated under an arrangement in which the station supplied a hut, the ground, water, tools and basic rations, for which they supplied the homestead and barracks with vegetables free of charge, and made pocket money from sales to other station families. They were probably stragglers from the later gold rushes and disappeared, who knows where, during the 1930s. A few years ago you could still find the remains of their unlined tin huts, the walls plastered with Chinese newspapers, around some of the old stations.

Keeping up standards was an important social issue in the 1920s, even away from the big stations. Jill Ker Conway captured this attitude when describing her parents' life as soldier settlers in the back country near Ivanhoe in New South Wales: 'No standard of cleanliness or nicety of domestic arrangements was sacrificed because of the limited supply of water, the outdoor privy, and the never-ending red dust. By six in the evening the children were bathed, the struggling geraniums watered, the dinner table set with starched linen and well-polished silver, and she was dressed immaculately for dinner. My father, freshly shaven and dressed for leisure, would join her, and they would dine as though there were a cook and a maid in the kitchen'.[4]

But not all station owners, even successful ones, went to so much trouble. In March 1922 a correspondent visited Dawson's station near Hughenden in north-west Queensland: 'Dawson is somewhat of a character. He has a small block, I think 40,000 acres of good country, with a bore and a dam, and he usually has a sheep to the acre. He lives in an iron humpy – alone. Employs no-one, rides about among his sheep, and if he sees one with a big lot of wool, or looking flyblown, the dog catches it and Dawson shears it, puts the wool in a tree fork, and comes around with a cart later and gathers up these fleeces. And so Dawson goes on accumulating, and his success often induces the remark "I'm going to manage like Dawson". But they couldn't live like Dawson. They'd want a cook, and housekeeper and cows' milk and butter and lots of things. This kind of life is all right so long as it lasts for those who like it, but the end usually comes at the first illness. He lived hard – mutton and johnnycake and johnnycake and mutton and black tea, and when men live like that they've no inside fat, no strength, and the first illness ends it.'[5]

Homestead lighting was provided by an external gas plant for the substantial establishments, or 'Aladdin' kerosene pressure lamps – carbide lights that gave a wonderful white light but were liable to explode without warning – or the humble hurricane kerosene wick lamps. Fuel for cooking and heating was wood, and huge quantities were consumed – gathering and cutting was a major operation. Food was kept more or less fresh with Coolgardie drip safes. Even small towns had a freezing works, and blocks of ice were packed in sawdust and then sewn into hessian bags for weekly delivery on the mail truck. Sheep were killed daily and meat was eaten fresh and quickly, or made into curries that might last longer. Meat was also salted or pickled, and every station manager had his favourite recipe. People were less squeamish then about food that might be slightly off, and station workers usually carried sharp pocket-knives to slice away the less savoury parts of ageing meat. The canvas water bag, now made redundant by styrofoam coolers, was ubiquitous. The standard rectangular bag with its china spout and wire handle and the 'verandah' tubular model cooled water through the action of evaporation on the moist surface. Once you became used to the baggy taste the water was wonderfully cool.

Homestead comforts began to improve around mid-decade. In 1926 the Hume Pipe Company was offering 'city sanitation for the country home' at £30, claiming: 'with a Hume septic tank installed you have a complete sewage system on your premises, allowing for the installation of flush closets. Fly dangers, typhoid, all deadly risks of contamination are eliminated.' That year Moffat-Virtue was offering their Phelps 32-volt electric lighting plant. The Isko Distributing Company was selling an electric freezing machine to 'make a daily supply of ice cubes for table use or sick room from pure drinking water'. Household refrigerators, powered by kerosene, were not to come into general use for another 20 years. In 1926 at *Culingral*, near Merriwa, Mr Thomas B MacFarlane had his homestead well set up – 'lighted throughout by electricity, and with a hot-water service to the various bathrooms, and mosquito and fly-proof gauze to windows, doors and verandah openings – its conveniences are most complete'.[6]

Social life on the stations was closely associated with sport. The bigger operations had cricket teams that played neighbouring stations on an irregular basis. As a child in the late 1940s I discovered an ancient and mouldy cricket kit that had been used in the early 1920s at our home, *Buckiinguy*, near Warren. The chipped bats were the colour of old mahogany and signed by those Edwardian heroes Warren Bardsley and Clem Hill, the open-work pads were made from cane lathes, and the ball was stitched with copper wire for use on ant-bed wickets. Rowand Jameson wrote of a jackaroo's life at *Westlands* near Longreach in the early 1920s: 'Our recreations were mainly tennis on Sunday if we were not working and at times cricket. At night we usually read or listened to the gramophone, with occasional boxing bouts. Arrilalah and Isisford races gave us an opportunity to nominate station horses, and most of us rode our choices.'[7] Local race meetings were big gatherings. In 1926 the *Murgoo* picnic races in outback Western Australia drew a crowd of 150 whites and 400 Aborigines for the two days of racing and three nights of dancing in the wool room. Visitors provided their own swags and their own supplies of petrol and oil. The shearers' quarters were walled in with hessian for the ladies, and the men camped in tents or the woolshed catching pens. Any person overdoing things with liquor was arrested and manacled to a tree for the night, and the following morning might be sentenced to leave the station within 10 minutes. The *Murgoo* cup was a trophy worth £35; other races had stakes of £25 cash and an iced and decorated fruitcake. At bush race-meeting dances, guests might perform the new 'shimmy', which shocked the older generation and prompted a clergyman to write: 'If these up-to-date dances are within a hundred miles of what I hear about them, I should say the morals of the pigsty would be respectable in comparison.'[8]

The stations took great pride in the quality of their horses, and strong passions could be roused. There was an amateur race meeting in Blackall during the 1920s which amounted to a match between the *Northampton Downs* and *Terrick Terrick* horses. Blackstock, manager of *Northampton Downs*, was judge, and when his decision on the winner of the main race was disputed, a wild brawl erupted between the jackaroos of the respective properties. My father, then overseer at *Terrick Terrick*, was seen in the middle of the melee swinging away and having a wonderful time.

At the other end of the social scale were graziers such as Otway Falkiner, who maintained a stable of racehorses in Sydney. He was reputed to have won £40,000 when his horse, David, won the Sydney Cup. In October 1920 he wrote to his betting agent, J Hackett of Tattersalls Club: 'I am enclosing a cheque for £2760, and would be obliged if, after deducting 500 for yourself, you would attend to the payment of the undermentioned amounts for me: L Levy 250, Vandenburg 100, Joe Browne 1000, Sid Baker 910.' When placing bets Falkiner communicated with his agent through coded telegrams, to shield his intentions from listeners-in at the party-line telephone exchanges.

Men's dress had changed little since before the war, and suits were obligatory for most social events. Every town had a tailor, and it was considered poor form to wear 'off the hook' kit. Those less particular could buy a single-breasted woollen worsted suit from Murdoch's of Park Street, Sydney, for 84 shillings. Women's fashion, on the other hand, had changed profoundly. The tubular look of the 1920s originated in Europe, where food rationing and the need for women to work in previously male-dominated occupations resulted in their losing considerable weight. With the demise of corsets, out went the classic 'hour glass' figure. Wartime work also meant that women had less time to spend arranging their hair, so short hair became the order. High heels, once associated with Paris and immorality, now came into general use, though conservative sections of the medical profession condemned them as causing uterine displacement and being a threat to the birth rate.

In the early 1920s, large gatherings of people were discouraged, for fear of the Spanish influenza that had entered Australia with returning troops. Before antibiotics became available, it killed more than 50 million people around the world. There were countless small tragedies; my father was engaged to a Lockhart girl, Nelly Drummond, who was tragically carried off at 18 just before they married. There was even a move to suspend Sydney ram sales in 1921 because of the 'wave of germs and bad air which meets the buyers on their arrival at the entrance'.

Prior to the 1960s most stations were dry camps – only the manager was allowed to have alcohol. But Christmas was different, and *Terrick Terrick* floated buoyantly on a dense cloud of alcohol. The day began at 7 am with the manager starting a round of visits to every person living in the homestead area, carrying a tribute of bottled beer and sharing a drink with each. It was customary for the beer to be one of the Melbourne brands, which were far superior to local brews. There was Christmas lunch in the jackaroo's barracks, usually in the English style of a roast and vegetables followed by plum pudding and hot custard – despite the searing heat that usually came with Christmas – and accompanied by more beer. Drinking wine was unheard of, and would have been considered unmanly.

The afternoon was spent sleeping off lunch or visiting the station hands and drinking more beer, because the usual social divisions were suspended for the day. Tea was in the homestead, to clean up the leftover ham and turkey and drink more beer. Then, in my time, there was a 16 mm picture show held on the lawn in front of the barracks – usually some old black-and-white musical or, if we were lucky, one of the latest Ealing Studios' comedies. The audience greeted the first reel with gales of laughter. They quietened down during the second, and by the third most were asleep. Towards the end, the soundtrack was accompanied by the sound of people heaving into the hedge and groaning with alcoholic remorse.

There was always a Santa arriving by buggy with a small present for everyone, but the choice of Santa was often unsound. At Kynuna in Queensland in 1923: 'Old Scotty fell for the job – he was just the man, a stranger with a deep gruff voice. Scotty was taken across the road to Absalom's Hotel to change into his regalia. He had not been long changing in the hotel, but it was quite evident that he must have disposed of two or three "rosiners" in the process, and he was walking a bit high. Everything was going well. With children gathered round him, Scotty was patting them on the head and handing out his presents when Fred Craig from *Mimong* walked past and gave Santa's false beard a friendly pull. This was too much for Scotty, who was always cranky, and he said: "If you do that again I'll give you a smack on the bloody kisser". The looks of astonishment on the youngsters' faces would be hard to describe. Instead of crowding round for a pat on the head they were inclined to stand off a bit – Santa was not the nice old man he was cracked up to be."[9]

Moving pictures had arrived as early as the 1920s – Charlie Chaplin, and westerns such as *Riders of the Purple Sage* and those of Hoot Gibson and William S Hart. Of course, these were silent movies, and the style of acting was distinctive: 'The early film star usually grimaced at his audience like someone trying to convey news of terrific importance to a stone-deaf and half-witted child.'[10] In 1925, when manager of *Cambridge Downs*, my father took his Aboriginal stockmen to see a Tom Mix western in the town of Richmond, north Queensland. When the hero galloped towards the audience blazing away with his six-shooters he awakened memories that were all too recent, and they took off into the scrub, not to be found for days.

Reading was the most important leisure activity, and the popular lending libraries sent out monthly parcels with titles by Nat Gould, Sax Rohmer, O.Henry, Edgar Wallace and PG Wodehouse. Children, strangely, read comics about a romanticised English public school life circa 1910 – *Champion, Wizard, Gem* and *Magnet.* Magazines such as *The Illustrated London News* were popular, and nearly everyone had the daily metropolitan newspapers sent out in the mail, even if they were a week old when they arrived.

Reaffirming Australia's close traditional ties to England, the popular Prince of Wales toured in 1920, visiting several stations including *Wingadee* at Coonamble and *Canonbar* at Nyngan. Newspapers gushed about his riding, kangaroo and emu hunting, and drinking black tea from a tin pannikin. It was claimed that at *Wingadee* he was given three hearty cheers by 30 mounted station men, but, as most would have had fresh memories of the bloody shambles on the Somme or the Salient, the report may have been hyperbole. If the many rumours about the Prince's social life on tour, and the number of 'out-of-season lambs' he begat, had any foundation, then there were many Australians in their eighties at the turn of the 21st century who carried royal blood.

All in all, things were getting better in the bush and life easier, particularly for established woolgrowers. Unity Pheils recalls her childhood at *Coree*, near Jerilderie, in 1926: 'For many years we had escaped the heat of summer by spending some weeks, including Christmas, in Melbourne. We caught the train in Jerilderie … the journey took several hours, and we travelled with much luggage and several servants. In 1926 there had been a series of good seasons with high wool prices, there was money in the bank, and I remember various purchases, including a yacht, being discussed. My father did buy an imposing residence called 'Stanhope' in Toorak Road, South Yarra … the house was beautifully furnished and decorated. Outside there was a sizeable garden with a hard tennis court and stables. My mother enjoyed entertaining the cream of Melbourne Society to grand dinner parties there.'[11]

Wool growing was a profitable business for most of the 1920s, thanks to surging post-war demand for clothes from world consumers who had been starved of quality goods for five years. Growers were content, and there were few signs of the savage tribal political warfare that was to break out in their ranks during more troubled decades. Many shared the feelings of returned men everywhere: 'To be able to sleep all night on a spring mattress, to have the company of women and children, to be done with mud and trench-rats. To be given something else to eat than bully-beef, biscuit, and plum-and-apple jam, and above all to be masters of their spare time – such relief made them care very little about what was going on in the public way.'[12]

CHAPTER THREE

THE THREADBARE THIRTIES

We've cracked our jokes and had our fun
And done our share of toiling:
We hope those mates who follow on
Will keep the billy boiling.

JACK MOSES, *KEEP THE BILLY BOILING*

As the Great Depression began to bite, wool prices dropped away, woolgrowers fell on hard times, and the buoyant confidence they had felt since the end of the war evaporated.

By 1932 around 800,000 people, or one in eight of the entire Australian population, were out of work and receiving government handouts. Employees of the big pastoral companies lived in fear for their jobs. My father had his first manager's appointment to *Buckiinguy*, near Nyngan in New South Wales, in 1926, and was married in 1927. He and my mother were deeply concerned that the Australian Estates Company, which was undergoing a restructure at the time, would put him off or sell the station. Stock agent Alec Watson, a charming man who I knew in the 1950s when he had become one of the most trusted and popular agents in the business, had just been appointed manager for the wool firm New Zealand Loan in Jerilderie. He later recalled: 'My first and lasting memory was when I got a letter from head office saying that my salary was cut by about half. This was a sad blow, but it was a case of accepting that or of joining the masses of men walking the roads.'[1]

Men, as they had done before in times of depression, took to the roads in thousands. Some were on foot, some pushing (seldom riding) bicycles loaded up with swag roll, the essential water bag, billy can, hessian bags for meat and calico for tea, sugar and flour. Some genuinely sought work, others wore the fixed stare of those who had given up hope, and for whom constant shifting had become an end in itself. They turned up at the stations day and night, weary, unwashed, their clothes reeking of smoke.

James Mitchell, manager of *Wellshot*, at Longreach, recalled the times: 'There was never any shortage if you were looking for a man, there were always bagmen passing through. I allowed anyone who was carrying his swag to camp at the shearing shed for two days, and sometimes there would be seven or eight bagmen there at any one time. On the third day we would move them on after giving them some rations and a plug of tobacco. They would walk on to Portland and do the same thing there.'[2] Reg Larkin, a grazier from Ben Bullen, near Lithgow in New South Wales, could remember whole families sitting out on nearby rocks warming themselves in the morning sun. They were out-of-work tradesmen and small business operators who had moved from Sydney in the hope of panning a little gold from the local streams. They had no tents, and lived in the scrub under brush shelters. Though on short rations himself, Larkin helped out with a few potatoes and a little meat, and in later years recalled with pleasure delivering telegrams containing offers of work, which were addressed 'So and So, on the creek near Ben Bullen'.

TH Hunter, then owner of *Evora* at Blackall, was known far and wide as a friend of the swagman. 'Any swaggie on the road was welcome to camp in the shearers' quarters and was provided with meat, flour, tea and sugar; if a smoker, with a small ration of tobacco; and if down and out, with boots, trousers and a shirt, all from the station store. It was rather pathetic at this time to see so many men on the roads with their swags up; many were quite young and were quite genuinely looking for work. The practice at that time was for them to draw rations through the police and then they had to move on to another town for the next rations.'[3]

Others were less charitable than Mitchell, Larkin and Hunter. One Riverina pastoralist would only offer rations in exchange for some form of work. If there was no productive station work available he would set the swagmen to shifting great piles of sand from one place to another, using excessively large shovels made expressly for the purpose.

Things were tough around Longreach: 'The Depression years were terrible. Credit was stopped and there was just no money around … there were lots of people who had to give their shopping lists to the bank manager before they could buy anything in town. If they had a luxury like marmalade on the list it was crossed out and plum jam was written down instead. I can remember riding up to town to do some business and then riding home again before lunch because I didn't have enough cash to buy a meal at the café.'[4]

In the late 1960s I classed sheep for a family in central-western New South Wales who 37 years earlier had suffered the indignity of having to present their monthly stores list to the local wool firm manager for approval. Though they acknowledged that the discipline had saved the farm for their children, and that they still dealt with the same wool firm, their deep bitterness had never evaporated. Their shame was that everyone in town knew about their plight.

Many of the banks employed pastoral inspectors, who were often former pastoralists, to visit their most indebted customers and report on their status and standard of management. They could be tough. One New South Wales pastoral inspector wrote to his head office regarding a client: 'I have told Lack that I didn't think he had been as economical as he might have been. I noticed that there is a piano in the house, but in going through the accounts could not see how it was paid for.'[5]

It was probably in those times that the famous story about rural bankers was created: 'One of the local bank managers died of heart attack. His death was widely reported in the local papers, which confirmed that, contrary to the district views about him, his parents were in fact married. That afternoon a local grazier called at the branch and asked for the manager. "I am very sorry," said the girl at the front desk. "I'm afraid he died yesterday." Next day the grazier was back, and again asked for the manager. This was repeated three times over three days. The third time the girl at the desk gave her standard reply, and asked, "Aren't you the gentleman that came in yesterday, and the day before that, and the day before that?" The grazier replied, "Ah yes, and I am going to be back tomorrow and the day after. I want to hear again and again the good news that the bastard is dead".'[6]

Generally, graziers suffered less deprivation than unemployed workmen, because as a last resort they could be self-sufficient and live off their own production, however boring and inadequate the diet of meat and damper might be. But to many, the Depression was a sharp cultural shock. Unity Pheils recalled the change in circumstances at *Coree*: 'The family was living graciously, surrounded by a large staff of servants, with no financial worries, and a busy social life at *Coree* and in Melbourne … [her father] was ill-prepared financially for the downturn which came in 1928. He wrote to the manager at the Bank of Australia in Jerilderie asking for an increase in his overdraft limit by £10,000. Both he and Una [his wife] had become accustomed to an extravagant lifestyle which neither of them was prepared to modify. In 1931 the bank refused to provide further credit to cover *Coree's* running expenses.' The family made some token efforts to trim costs: 'The cook had no kitchen maid. The laundress went, the sheets were sent out and everyone did their own washing … there was only one housemaid and it was help yourself at meal times … the gardeners went.' [7]

Eventually the McCaugheys (Unity Phiels' family) decided it would be cheaper, more congenial and better for the education of their children to live in the south of France. So they departed *Coree* with a nanny and 54 pieces of luggage.

In October 1930, Joe Thomson, the competent manager of *Boonoke*, who many believe underpinned the whole Falkiner family empire, offered to serve at a reduced salary, but his offer was refused. However in 1931 all managers' salaries were cut by 10 per cent, though they were restored when the wool market improved in 1934. Some could not stand the strain. In 1931 the Falkiner directors noted that 'J Doherty, managing overseer at *Zara*, had broken down under the strain and his condition was serious, we had him in Melbourne under treatment and his doctors had some hope of his recovery'.[8]

Surprisingly, Australian sheep numbers held steady during the 1930s at around 110 million, despite the wry reflection on the current feeling about sheep made by a Riverina woolgrower in 1934: 'Sheep are born in the autumn, mortgaged in winter, and put on the road to die in the summer. The prices of sheep are determined at Newmarket, and go up when you have sold and down when you have bought. An agent working for a country firm was sent to Newmarket recently to watch the sheep sales, and after a few days' deliberation he wired to this effect: "Some think they will go down and some think they will go up. I do too. Whatever you do will be wrong. Act at once".'[9]

In 1933 there were 6.6 million people in Australia, and 2.4 million of them lived on farms and stations. Domestic life in their homesteads and cottages, before the days of washing machines, dishwashers, vacuum cleaners, detergents and effective cleaning fluids, was a routine of drudgery. But, for those who could afford them, additional comforts were slowly becoming available. In 1930 Adams and Co. in Brisbane were advertising the Gnome refrigerator: 'Does not need any power. Just place a primus stove under the Solution Chamber for about an hour and a half. This heat, and the cooling water, starts refrigeration that lasts for 24 hours. It has proved itself by keeping meat fresh for 12 days in intense tropical heat.' By 1933 Anthony Hordern & Sons in Sydney were offering the Selfreezer in wax-finished maple cabinet for £37: ' … the freezing unit is suspended inside the cabinet connected by stout pipes to the chemical container, which hangs outside. Every morning the freezing unit is lifted out and heated as directed.'

More home lighting plants were being installed. In 1935 the Farm and Pastoral Supplies Company could deliver 110-volt lighting plants for £115 and 32-volt for £100. The problem was that very few household appliances were available that could utilise these voltages. Jameson, then manager at *Evora* near Blackall, described the changes: 'Later 32-volt electricity was installed, and to add to this a kerosene Electrolux refrigerator was sent up … It was indeed a luxury to come home to a cool drink on a hot summer's day, and to be able to switch on a light instead of carrying a lamp around.'[10] Amazing novelties were being made possible by electrical power. In the early 1930s journalist Lionel Krone visited *Bullagreen*, near Gilgandra, where the manager Beck 'fired up the big engine in the woolshed to start the electric light so that we might have a moving picture show. The films were the Cine Kodak pictures of station life and other things, including bathing beauties at Bondi and Manly. This latter film was easily the most popular amongst at least some of the station hands.'[11]

The basic station diet was still meat, bread and tomato sauce or pickles, but in 1933 a scheme was developed in Queensland for forwarding supplies of cheap fruit inland through the Country Women's Association, with the assistance of the railways and school teachers. Members of State Parliament were rather patronising when they 'gave the scheme their benediction, and outlined a proposal for sending small oranges and similar fruit not ordinarily saleable, but wholesome and cheap'.[12] The greengrocer in Longreach at that time was notorious for sending tainted fruit and vegetables to the stations at the end of the long mail runs in the certainty they would be unable to send them back.

Domestic help for the homesteads was, understandably, in short supply given isolation and lack of social amenities. In 1934 the Queensland Selectors' Association called for an immigration system by which girls would be assisted to come to Australia. Their press release could have been more felicitously worded: 'Provided suitable types are selected, members are prepared to guarantee service to the girls.'

In 1938 Levin and Co., of Kent Street, Sydney advertised an appliance which, for those who could afford it, was to transform the comfort and convenience of homestead kitchens. The Aga slow-combustion stove, run on coke, boasted a 'capacious oven that takes a 20-pound turkey, a well-mannered simmering plate, and design to harmonise with modern kitchen architecture'. Its huge advantage was that it provided constant heat, so cooks no longer had to get up long before dawn to rake out cold ashes, light the fire, and allow sufficient time to generate the heat to cook breakfast. The stoves also provided instant hot water and, because they radiated little heat into the room, kitchens were no longer turned into furnaces by wood fires during summer.

Though isolation was diminishing and services were improving, the bush could still be an uncertain and dangerous place. In May 1930 reports from Townsville advised: 'The mystery of the disappearance of Mr J Ezzett, an aged recluse, who owned *Bora* near Croydon, and who left the station on February 3, has been cleared up by the discovery of the body by mounted police and black trackers three miles from the homestead. From tracks, it appears Mr Ezzett left the homestead early to muster cattle, and his horse did not have a surcingle. As he was galloping after cattle the girth broke and he was thrown heavily to the ground. His horse disappeared, and after placing his saddle carefully against a tree, the deceased staggered for a hundred yards, where he collapsed and died. When discovered nothing remained but a heap of scattered bones and torn clothing. The bridled horse has not yet been recovered.'

Social life on the outback stations was still constrained by lack of mobility. Few employees could afford a car, and in any event most pastoral companies forbade private ownership of vehicles. Access to town was at the discretion of the manager in allowing use of station vehicles, which was restricted to a few times a year. Pastoral inspectors on their quarterly visits paid close attention to the condition of access roads so as to detect any excessive use for trips to town – managers wanted their staff on the job and available, not going to town or thinking about town. So people were thrown on to their own resources. Alf Walker recalled his time at *Beaconsfield*, Ilfracombe: 'We were not encouraged to leave the station at weekends, and in any case there was no transport, as private cars were a rarity. You filled in your time as best you could, just laying about, flirting with the housemaids, playing cards, or making fun like athletic contests or riding a buckjumper in the horse yards. It wasn't too bad if you were a reader, but not everyone was. We didn't see a great deal of Ilfracombe, but if we went in we had to see Miss O'Kane. Her beer came from a keg on the bar, covered by a wet bag. People say that her beer was always flat but we didn't notice or didn't care.'[13]

Pastoralists and managers usually took holidays during the peak of summer, and tended to choose the same destination each year. Most people from central-western New South Wales headed for the Sydney beach suburb of Manly, whose many permanent residents spent their summers in Katoomba and rented their homes for the season. The bars and the beaches were full of talk about sheep prices and wool. Some wealthy graziers might enjoy a stay at Sydney's Wentworth Hotel, which

Men took to the roads in their thousands during the Great Depression *(Strangman Collection, National Library of Australia).*

claimed to be the only establishment steam-heated in winter and air-cooled in summer, and where every room had a telephone and hot running water. Tariff was 12 shillings and sixpence bed and breakfast, or 20 shillings fully inclusive. At the top of the range they could choose the Australia Hotel in Castlereagh Street, where a single room with bath and toilet cost 17 shillings and sixpence, breakfast was three shillings and sixpence, light luncheon in the Wintergarden two shillings, and dinner a minimum charge of four shillings.

Dress codes were still formal, a suit being required for most occasions, preferably one run up by a tailor. My mother was mortified when father, called to Melbourne for a conference at Australian Estates head office, bought a suit off the hook with the then popular 'Oxford Bags' – wide with a pleated front. She feared that appearing in such racy dress could jeopardise his career prospects with an old and conservative English company. Work clothes were ordered from mail-order catalogues, or purchased from the hawkers who made regular trips around the bigger stations. The station would wake one morning, and they would be there – an old covered wagon, two hobbled horses nearby, and a black iron pot steaming over the fire. City bred or schooled lads who considered themselves sophisticated might scoff at their wares, but that night would gather round the wagon, smell the smell of new clothes, and end up buying something. James Mitchell, when manager at *Wellshot*, commented: 'The other itinerants we used to see were hawkers, especially during shearing time. It was quite an event when a hawker turned up. They sold just about everything, but most seemed to specialise in haberdashery. They would stop at the shearing shed and then go on to the outstation. I rather encouraged the hawkers because when they called there was one less reason for men to go to town.'[14]

Moleskin trousers were universal. Not the rather fashionable article we know, but great heavy things as thick as your thumb which would stand up in the corner unassisted, with the choice of grey or olive-green. The custom was to buy two pairs in winter, so that by summer constant washing had lightened, whitened and softened them. They came in two sizes only, large and larger; shapeless tubes which were pulled up to the required height and cinched tight with a leather belt hung round with accoutrements. These might include leather pouches containing pocket watch, knife, snakebite kit, and a tin box of waterproof wax matches. Smartly turned-out stockmen might add concertina leggings and Cuban-heeled riding boots, which could nearly cripple them during work in the sheep yards.

During the grim years of the early 1930s the real problem for woolgrowers was that they were caught in the pincers of a classic cost-price squeeze. Abroad, their fortunes depended upon the willingness and ability of consumers in the developed world to walk into shops and buy clothes made from wool. Competition for the consumers' attention was coming from new and cheaper synthetic fibres, radios, household gadgets and, during the Depression years, the fight for survival. At home, wages and living standards were supported by high tariffs on imported goods; Australia was a high-cost business environment.

The government's policy of closer settlement, which was splitting up the larger properties and creating over-capitalisation and fragmentation, was also making its contribution to rising production costs. The Queensland Land Administration Board found, in its 1929–30 annual report, that the costs of running an average-sized grazing property in central Queensland had doubled during the previous 15 years. In 1932 it was estimated that operating costs were around £18 per bale of wool. With the market only paying £11 a bale these woolgrowers were in trouble. Labour costs were the key component, and many stations were employing one man to 3000 sheep. By the late 1990s it was to be one man to 10,000 sheep or more. In 1934 *Barratta* supported a manager on £750 per annum (a generous salary for those times), an overseer on £220, and 26 other employees. These included a bookkeeper, horse driver, cook, three jackaroos, 15 station hands, housemaids, gardener, houseboy and two rabbiters. By way of contrast, in 1999 a manager, overseer, and three men ran the same area carrying the same number of sheep.

Many people were needed because stock work was done with horses, and horses absorbed a lot of time. In 1934 the plant list at *Barratta* showed that animal power still dominated. There was a Buick car, an Albion lorry and Chevrolet utility, and also two buggies, a wagonette, two bullock wagons and tackling, a horse wagon, trolley, three spring carts, three drays and four sulkies. To be sure, horses had advantages – the high viewing platform provided by a buggy or saddle meant close observation of pastures, and stock were never rushed or harried about. But real breakthroughs in reducing operating costs only came in the 1960s when horses were completely replaced by unromantic motorcycles and four-wheel-drive vehicles. The typical station routine was for horses currently in work to be run into the yards before breakfast, usually a jackaroo's job taking turns day about. It was a point of honour to have the horses yarded, finish breakfast, cut lunch, and be on time for orders at 7 am, sometimes a big task given the horse paddock might be 2000 hectares or more. The overseer allocated tasks at orders, then 10 or 20 men might catch horses, saddle up, pack their saddle bags, take the buck out of fractious animals, and set out in small teams. There was always a sense of high spirits, optimism and purpose among a group of young horsemen setting out for the day. There was companionship, pride in being astride a smart horse with smart gear, and the certainty that what they were doing was important. But an hour might have passed since orders and it might be another hour or more – alternately walking, trotting and cantering on the overseer's order – before the destination was reached and the job begun.

According to workload and quality of pastures, horses would stay on the working complement for a month or more before being turned out for a spell and a fresh team brought in. There was little sentimentality about horses when they were used as an instrument for work; they were treated hard – it was rare to see a fat horse on the stations. Each horse was named according to its breeding, habits, appearance, or vices. There might be Mistress, Countess, Duchess, Blaze, Missile, Twister, Dagwood, Rufus, Amity, Victor or Striker. Two or three horses were allocated to each man, to ride day about. The bringing in and allocation of fresh horses was an important day, eagerly anticipated by the good horsemen and dreaded by the less adept. Spelled horses were likely to have some fire in them – they might buck, rear, bite, strike or kick for a couple of days before settling into work. Most stations bred their own horses from 80 or more brood mares and a stallion or two – usually a thoroughbred of modest racing record but with the hardy constitution and walking ability needed for station work. Grooming, care of feet, and shoeing were time-consuming. Maintenance of saddles and gear was usually reserved for rainy days. Saddles, bridles, surcingles, girths, cruppers, breastplates and stirrup leathers were washed clean of caked sweat with warm soapy water and treated liberally with saddle dressing to keep them soft and pliable.

On most sheep stations all that remains of the old horse culture is likely to be the saddle room, now probably used for general storage. Walk into it and you will still smell history – an acrid but not unpleasant scent comprised of leather, horse sweat, old saddlecloths, dust, leather dressing and chaff. It remains embedded in the walls and rafters as a reminder of the past.

CHAPTER FOUR

WARTIME IN THE HOMESTEADS

Who live under the shadow of war,
What can I do that matters?

STEPHEN SPENDER (1909–1995)

At the outbreak of the Second World War wool auctions were abandoned. Wool was a strategic raw material used for military uniforms, and to avoid it falling into enemy hands the British Government, as it had done in the previous conflict, agreed to buy the entire Australian wool clip for the duration of the war. The negotiated price, though not excessive, was higher than that prevailing pre-war, and was sufficient to afford growers a reasonable return.

Wartime rationing and government controls over nearly all aspects of life in the bush were inconvenient but not oppressive, not least because most farms and stations were self-sufficient in regards to meat, milk, butter, eggs and vegetables. Domestic staff was unobtainable, so managers' and owners' wives found themselves cooking for station staff as well as doing general housework and laundry. This may not sound too burdensome to the current generation, but it needs to be remembered that the old homesteads were originally built on the assumption of a full complement of domestic staff. Kitchens were rambling and inconvenient, and often a long walk from dining rooms. Cooking utensils remained basic, with no labour-saving gadgets such as dishwashers or mixers. Travel in rural areas, though much improved, was still slow by current standards, so the large outback stations could expect many drop-in guests who often arrived without notice and needed to be fed and put up overnight.

There were few household appliances such as vacuum cleaners, and no detergents or efficient cleaning fluids. Housework involved Sunlight soap, wax, scrubbing brushes, methylated spirits, sandsoap and lots of hard work. Arrangements for laundering were often primitive. A station wife from near Longreach recalled: 'The laundry facilities were a bench in the open air with two laundry tubs and a scrubbing board. There was a hole in the dirt in the garden with bore water lying in it, and I would scoop water from this in a kerosene tin to fill the tubs.'[1] Many young women of the time had aspirations about being hostesses rather than drudges, and they found wartime work onerous. My mother, who had become a station manager's wife in 1927, looked on in wry amusement at the interest in domestic cooking that grew during the 1970s. She commented: 'In my time you learned the basics, and then worked hard at reaching a social position where you didn't have to cook.'

When post-war migration began to flow, there were calls to alleviate what the rural journals called 'the present conditions of unmitigated drudgery for country women'. One 1948 editorial, which showed little understanding of changing social attitudes, commented: 'The pastoral industry cannot flourish as it should unless it is based on the fundamentals of a comfortable home life. Australian domestic labour is apparently unobtainable. The only alternative is to obtain immigrants with a specific leaning towards such work. The Chinese are the obvious choice … they could be admitted under indentures. Failing them the most suitable type would be the Southern European.' An angry and intolerant correspondent to the rural papers agreed: 'One of the most important living amenities in the outback is a supply of green vegetables. For gardening, what is wrong with importing the old coolie Chinaman as both cook and gardener, especially seeing that we allow enemy aliens to infest farms at a rate of pay they could never even dream of in their own land?'[2]

In 1946 at the annual general meeting of the Western Australia Pastoralists' Association, President Lefroy made an ill-advised comment when he suggested 'the employment, under proper controls, of labour from the Empire, irrespective of race or colour. It has been proved fairly conclusively that white employees do not want domestic work. Consequently the employment of people who by nature are suited to the ruling climatic conditions would impose no hardship, but on the contrary make for better living conditions.'[3] Unsurprisingly, these moves provoked a devastating rebuke from the Labor Government's Immigration Minister Arthur Calwell. He directed his venom at the Federal Council of the Graziers' Association, which he claimed had approached him 'because they wanted Chinese cooks for outback stations. When I told them they could not have Chinese cooks, they said they were prepared to take Italians.' He called them 'cynical and greedy capitalists who want a return to the good old days, a return to Kanaka labour'. He went on in fine style: 'The ultra-conservatives and land barons would like vast pools of near-slave labour to make them richer. The blackbirders of the South Seas have gone, but their spirit lives on among the cattle kings and sheep barons, the money lords and the princes of profit. There are many who resent the independence of the Australian working man. When the spiritual descendants of the blackbirders turn their eyes to certain countries in which there is still coolie labour they go green with envy. In such a society they see themselves as superior to even the whip-cracking, lazy, gin-drinking overseers.'[4]

Food rationing was imposed early in the war, and was controlled through the issue of coupons, which were collected when buying rationed goods. Penalties for breach of rationing laws were severe – there was no limit placed on fines or prison terms, and in theory the authorities could lock people up and throw away the key. Food controls bred a huge and flourishing bureaucracy. Like something from George Orwell, there was a Director General of Agriculture, Controller-General of Foods, a Food Executive of four ministers, Director of Food Manufacture, Director of Foodstuffs, Director of Civil Supplies, a Commonwealth Food Purchasing Board, Food Specification Committee, and Controllers for meat, egg and tea supplies. Sugar was rationed from September 1942 to one pound per person per week, and tea from April 1942 at one ounce, later lifted to two ounces following protests. Butter was limited to eight ounces from June 1943, then reduced to six ounces in 1944, so people returned to the old stand-by of spreading their sandwiches with dripping. The meat ration, though not a problem in the bush, was 2.25 pounds per week. Australia continued to live under rationing post-war as food supplies were diverted to a hungry Britain. We were consuming 77,000 tonnes of butter under rationing, and exporting 72,500 to Britain, and in 1946 the Country Women's Association organised shipment of 4000 four-gallon tins of fat.

Though relaxed progressively after the war, rationing and food restrictions in Australia were not fully lifted until 1950.

It was often said that the bush ran on tea, sugar and tobacco, and the rationing of these items was seen as striking at the very heart of bush culture. Station people drank gallons of scalding tea even in the hottest weather – one northern pastoral worker protested to the Controller for Tea that he needed 15 pannikins a day to keep going through the summer. Special arrangements were put in place by no less than the Deputy Director of Rationing for extra supplies of tea, sugar and tobacco for shearing teams. Non-smokers were the exception then, and tobacco shortages caused much vexation. Station stores issued the monthly tobacco ration, and augmented supplies of the better-known brands with little-known local products. These were unpopular, and unused stocks lay in the stores until the 1950s. As a young jackaroo at *Terrick Terrick* in 1953 I was determined to start smoking, and the bookkeeper, who was a reformed smoker and had control over tobacco issue, was equally determined that I should not. So when my workmates got their issues of the popular brands such as Log Cabin, Havelock, Champion or Capstan, I received tins of blackened and dried-out weed sold under the names Mareeba or NQ out of old stock; stuff which caused even hardened addicts to cough and splutter. Sadly, I persisted and eventually won the argument.

Commercial life was severely restricted under regulations announced by Prime Minister Curtin in 1942. Sales and investment of capital except by permission of the Government were prohibited, and prices of all goods and services were pegged at 1942 levels. Profits, exclusive of tax, were pegged at a maximum of four per cent on capital; profits in excess were either taken in taxation or passed on through lower prices. The minister controlled all transfers of labour, and absenteeism became illegal. Speculation in commodities, such as forward dealing in foodstuffs, was also banned. Many of these controls were to remain in place for years; controls on prices of rural land were not lifted until 1949.

Clothing rationing was introduced in 1942 and, like food, the controls were implemented through a coupon system – 112 coupons per person per year. A woman's frock required 12 coupons; men's sports trousers 10. Peapes, near Wynyard Station in Sydney, offered made-to-measure two-piece suits, cut in the short 'Austerity' style, in Australian worsted at seven guineas and English at 12 guineas, each taking 30 coupons. Wool herringbone tweed working trousers were 30 shillings and five coupons. Clothes were patched and repatched; as small boys we were acutely embarrassed walking around with calico patches made from flour bags on our pants.

Petrol rationing imposed a limit of about 130 kilometres a week on cars for private use, and everyone was terribly busy because of reduced staff numbers, so district social events such as shows, picnic races and sports meetings were closed down for the duration. There was great excitement in the Yass district when, after a gap of many years, amateur races and a ball were held in the autumn of 1946. Though the Duke and Duchess of Gloucester accepted the invitation of President Mr E Merriman to attend, an observer noted: 'Cars were mostly old, with canvas showing through the tyres, and the clothes of the men were pre-war and shabby.' I can remember clearly the intense excitement I felt as a 10-year-old at the prospect of going to the first Nyngan show held after the war in 1945, particularly about visiting the sideshows and boxing tent. I can also remember clearly the bitter disappointment when we had two punctures in our rickety 1936 Ford utility and didn't make it to town.

Liquor played a big part in bush social life then, more so than it does now, and the war years were hard times for those who liked a drink. Beer was well nigh unprocurable, and home brews of various degrees of lethality and explosive powers were popular, though schnapps was, for some reason, readily available. Scotch whiskey was very rare, and a local product named Corio attempted to fill the gap. In post-war years, Corio was to become a reasonably civil drink, but in its early forms it would cause even hardened drinkers a nasty start. In those days it was customary to seal livestock deals with a 'spot', and when my father sold a big line of wethers to two local buyers and their agent during the late 1940s he offered them a Corio, with profuse apologies that he had nothing better on the shelf. I was just at that age when you see and hear everything, and knew exactly where he had planted a choice bottle of McCallums scotch, supplied per the good graces of Goldsbrough Mort. So, being anxious to please, I fetched and presented it to the group to help my dad out of his difficulties. They eagerly fell upon it and finished it off. Family relationships were strained for some time afterwards. Drinkers of those times could never quite believe their good fortune when quality liquor became readily available again, and for years they could be seen savouring each drop as if it were their last.

We lived a long way from the war, and we wartime children of the bush felt little sense of deprivation. There was enough to eat and enough to do. There was intense excitement when city cousins arrived to stay following the midget submarine attack on Sydney. There were occasional visits from former employees on leave – one gave me his army water bottle, which was a great thrill. Some neighbours had sons killed in action, but we saw them so seldom that the sense of loss was not transferred to us. We listened to the radio on old cabinet sets powered by wet-cell batteries, from which you were lucky to pick up two stations through the static. Young as we were, we listened avidly to the news commentaries, including 'The Watchman' on commercial radio, and William Winter, speaking out of San Francisco, on the ABC. But the ABC Children's Hour was the highlight of our day; we were all members of the Argonauts' Club, and achieving their 'Golden Fleece' or 'Dragon's Tooth' awards was cause for household celebration. The subjects of the childish drawings and compositions we submitted were not farm animals, bush picnics, or nature study but guns and tanks and battleships and Buna and Gona and Balikpapan. We worked through our weekly lessons from Blackfriars Correspondence School, and were thrilled to sometimes meet

Rations books with coupons for staples issued during World War II *(News Limited)*.

our invisible teachers on annual train trips to Sydney. We devoured the monthly parcel of books that arrived in tarred paper from the Bookstall Library. We received the daily papers once a week, and carefully arranged them in chronological order for reading.

The rare trips to town were great occasions, mainly because we could visit the cafe, usually run by a Greek called Theo or Con. There, despite rationing, we could take a break from the eternal mutton and stale bread, feasting on T-bone steak or fried flounder, accompanied by a milkshake or orange cordial so cold it snapped your tonsils, followed by ice-cream sundae. A few icon survivors of those cafes are still around, most famously in the towns of Gundagai, Yass and Brewarrina, and no child of the 1930s or 1940s can view their laminated booths, fluted milkshake glasses, sundae dishes, or crude murals depicting the Acropolis without a thrill of memory.

There were no school buses or local high schools that went through to matriculation level, so eventually we were all bundled off to boarding school in the city, aged anywhere from seven to 12. There we developed classic dual personalities, one for school use and one for home, and we learned about the world and about other people. Being thrown into a group of other children could be a cultural shock – there were many like myself who, because of isolation, had never dealt with anyone of their own age until they were 10.

With so many men in the forces there was plenty of work available on stations. In the isolated inland particularly there was a procession of shabby, shifty-eyed men – mostly calling themselves Jones, Brown or Smith – seeking refuge and taking up work of all kinds. They were dodging call-up, military police, manpower authorities, wives or alimony payments. Isolated locations were important to them because they could receive basic rations without the need for coupons, which required proof of identity. They worked hard, kept a low profile, made no trouble, had no desire to visit town, and preferred to be paid in cash. The stolid and stockily built Bob Johnstone arrived at *Buckiinguy* unheralded, on a bicycle with his swag up. He took on the groom's job, which involved butchering, milking, separating cream, making butter, and cutting wood. He stayed for two years and hardly spoke to anyone, then, without warning, without rancour, without explanation, rolled his swag, collected his old bike from the station store and was gone forever.

Downy-faced jackaroos came and went, often so quickly we hardly got to know them. Most seemed to be filling in time before assignment to basic flying training schools. A more exotic arrival was a sophisticated French-Canadian named Peter, who worked around the house. He had been in the French Foreign Legion, and had escaped from Indo-China to Australia in a small boat. You can imagine the fascination he held for a small boy who had just read *Beau Geste*. Peter became close to the family, and when he left *Buckiinguy* used to visit our grandmother in Sydney. After sending a few letters, he disappeared from our lives.

By 1942 Australian rural industry had lost around 25 per cent of its official workforce to the war effort. Single employed station hands were exempted or had their military call-up deferred; later this was extended to include working principals and managers in charge of three men or more. As the military situation deteriorated, and Australia came under serious threat, the numbers enlisting grew. Like their fathers in 1915 many were anxious to fight, and it was easy to resign from a protected occupation, or give false information; recruiting officers turned a blind eye. The Cunnamulla district was typical – by December 1942 more than 1000 men had gone from the local grazing industry. That year JR Baker, the chairman of wool brokers Elder Smith and Co, reported that of their total male staff of 682 at the outbreak of war, 329 (almost 50 per cent) had enlisted, and at that stage seven had been killed, and another 20 were missing or prisoners of war. In 1941 Otway Falkiner told his directors at *Boonoke* that a large proportion of the company's eligible employees had joined up. The board resolved that all were to receive from the company 'such a weekly sum as was necessary to make up, added to their military pay, an amount equal to their weekly wages prior to enlistment, and that their jobs be kept open for them after the war'. Some young men of military age, working in protected occupations but anxious to enlist, failed to get the required approval from parents or employers. Their situation in post-war rural society was often distressing, with the rest of their age group bonded by shared experience and the local gossips spreading innuendo.

Casualties and personal tragedies mounted. Up to October 1944 army losses alone amounted to 10,266 killed, 3335 missing, 25,458 prisoners of war, and 30,901 wounded. Worse news was to come when the full extent of losses in Japanese POW camps was revealed as the war ended. As had happened in the First World War, many of the brightest and best were lost, men who would have been at the height of their capacity when the wool industry hit the turbulence of the 1960s and 1970s. They included: Flying Officer John Falkiner, the promising youngest son of Otway, killed at 24 in a flying accident in England; Lieutenant Fred Treweeke, owner of *Nocoleche* at Bourke, killed in action October 1942, aged 30; Captain Ian McMaster, energetic only son of Sir Frederick McMaster of *Dalkeith*, Cassilis, died from wounds in Egypt 1942, aged 24; and the handsome Flying Officer Charles Fairbairn of *Banongil*, Skipton, shot down over Germany in 1945, aged 20.

Some of those left at home joined the Volunteer Defence Corps – Australia's 'Dad's Army' – and rushed about the country firing off ancient Lee-Metford rifles of Boer War vintage and having a lovely time blowing things up with gelignite. Most station people were too busy to participate, but felt it keenly when all repeating rifles were required to be handed in to police in case of invasion, and they lost their old 32/20 and .44 Winchesters. Some, anxious to defend their homestead if necessary, hid caches of arms, ammunition and petrol on their runs.

Labour shortages caused problems in getting flocks shorn. Traditionally there had been little seasonal flow or pattern to shearing, and there might be a variety of shearing times in any district. So in 1942 National Security (Shearing of Sheep) regulations were gazetted. Under these the Director-General of Manpower divided New South Wales and Victoria into eight zones in which shearing operations involving flocks of more than 1000 had to be carried out within a specified period. For example, zone one was all the area west of the Barwon and Darling Rivers from Mungindi through to Menindee, then across the South Australian border adjacent to Broken Hill. In this zone no shearing could be undertaken between July 6 and October 4. Zoning was not imposed in Queensland or Western Australia where there was already a strong seasonal and regional pattern established. The usual severe penalties, including jail sentences, could be imposed upon those breaking the regulations, which were not repealed until the 1946–47 season.

Pressure of work, drought and shortages took their toll on the middle-aged men who remained during the war. Station managers took their work very seriously; indeed today most would be classed as workaholics. Their days were long, with 6.30 am starts, or perhaps 4 am at stress times such as shearing and lamb marking, followed by a long stint in the office after dinner, attending to administration and the various returns. My father was running *Buckiinguy* with no overseer, no bookkeeper and five men where there once had been 16. He was coping with drought management of a 35,000 sheep flock, keeping a full set of station accounts, and also doing the books at *Raby* 50 kilometres towards Warren. The terrible summer of 1944–45 saw him hospitalised through a combination of nervous exhaustion, heat stroke and, probably, the effects of Corio whiskey.

Not all were so diligent. One famous Queensland manager issued early orders at 7 am from behind a large office window, dressed in tweed sports coat and tie. Little did his men know that underneath he was still wearing his flannel pyjama trousers, and once they were safely out of the way he returned to bed until what he considered a civilised hour.

Some were able to compensate. In 1942 the directors of F.S.Falkiner & Sons decided that 'in view of the failure of dividends for the last two years and the increased burdens thrown upon them by existing conditions, that the salaries and remuneration of executive directors be increased'. Otway Falkiner's package was increased by £1000 to a princely £5000. His son OMcL ('Mac') Falkiner scored an increase of £900, taking him to £2000 per annum. By way of comparison, at that time a competent company station manager was earning about £500 per annum.

The manpower situation was alleviated somewhat by the appearance in 1945 of new generation chemicals for control of blowflies and intestinal worms of sheep. The treatments available until then were unsophisticated and ineffective; anything that could not be cured by application of arsenic, bluestone, kerosene, nicotine, Stockholm Tar or turpentine was considered incurable. The new insecticides DDT, BHC and Gammexane provided lasting protection against blowflies and sheep lice, and the worm drench Phenothiazine was the first really effective antidote for intestinal worms. Many of these new chemicals carried with them serious health risks. They were used in diluted form and required precise and accurate measurements, whereas the old remedies were concentrated, mixed using crude rule-of-thumb judgement, and produced rich, creamy-looking solutions. The old station men had difficulty coming to terms with the new technology, and in their judgement the new chemicals looked anaemic and weak when mixed. I often saw them wash the solutions round suspiciously, paddle them with their hands, sniff and even taste them. If in doubt they would tip in some more concentrate to produce a nice rich-looking 'jollop'. It is a wonder that thousands of sheep and dozens of men were not killed outright.

CHAPTER FIVE

THE FABULOUS FIFTIES

In the day of prosperity be joyful ...

ECCLESIASTES

One day in March 1951 two middle-aged bachelors from the Coonamble district, shabbily dressed, wandered into the glossy showroom of York Motors, agents for Bentley and Rolls Royce cars in William Street, Sydney. One was eating a hamburger out of a brown paper bag. After convincing the supercilious young salesman they meant business, they decided there and then to take a gleaming black and yellow Bentley standing on the showroom floor, and one brother pulled out a cheque book. The other stopped him. 'You got the hamburgers for lunch so I'll get this.'

That same afternoon a western New South Wales grazier, ashen faced, staggered out into Macquarie Place Sydney from the nearby wool auction rooms and flopped onto a bench, head between his hands. A concerned friend who was passing walked over to see what the problem was. 'I have just seen my wool make a pound a pound,' was the shocked reply. All three had sold their wool at the peak of the hottest market the industry has ever experienced. During 1950 the Korean War had escalated, and there were fears that it could spread further to become World War Three. The US was making a firm stand against world Communism. England and the other Western Bloc nations were deeply in debt, their economies ravaged by the cost of fighting the Second World War, so the US, as the world's richest nation, would carry the burden of war in Korea and any wider conflagration that might result. The US Government was buying enormous quantities of woollen fabric for military uniforms and prices were high, but they were most concerned about future supplies should the war turn global. They began making tentative proposals for creating a strategic stockpile of wool by asking producing countries to allocate a proportion of their next wool clip under preferential conditions; locking up supply, not price, was their major consideration.

Negotiations were held on a government-to-government basis and in great secrecy. But as often happens when a number of people are concerned, particularly the truculent and faction-ridden woolgrower representative bodies that were consulted by the Australian government, information leaked and press stories began to appear. Though some of these articles were speculative, wool traders and processors in Europe sniffed the wind and reacted sharply. Many of these firms had been in the wool business for hundreds of years, and regarded themselves as the true custodians of the heritage of wool so, quite apart from commercial considerations, they tended to resent the entry of upstart new players from the New World. Fearing their access to wool might be disrupted, they joined a frantic rush to secure supplies at any price. Wool prices had been rising steadily since the late 1940s in response to strong demand for clothes from confident post-war consumers and the restocking of the near-empty processing pipeline. Then, between June and September 1950 prices jumped by 50 per cent, and began climbing towards the peak of March 1951, when they reached the dizzy average of £251 per bale. As a comparison, that was around six times the price being enjoyed by growers in the improved market of 2002, which was the best in more than a decade. Such prices had never been seen before, nor will they ever be seen again.

There was an added bonus. A wool stockpile of more than 10 million bales, most of it Australian, remained from the wartime purchase scheme set up by the British Government. This had been purchased at fairly modest prices, and was fed into the booming post-war market by a government body called the Joint Organisation. Substantial profits generated by the organisation were being returned to growers who had sold wool during the war years. The bush was awash with money.

It is common for later generations of woolgrowers, who lived through more stringent times, to reflect on the early 1950s as a period of great affluence, profligate spending and excessive behaviour. This was not the case. Those born around the turn of the 19th century, hence too old for service in the Second World War, had seen too many bad times to alter their frugal ways. Other than doing some renovations to rundown homesteads, taking a holiday in Manly, or adding a little Ford Prefect utility to their modest fleet, they went about their business as usual. Very few of them acquired villas in the south of France, stables of racehorses or luxury cruisers. Returned servicemen were too intent on rebuilding their lives to worry about riotous living. Those born in the late 1920s and early 1930s, so too young for the war, left school and entered their 20s in a more relaxed social and financial atmosphere than their parents had experienced. There was money around and consumer goods on which to spend it. Towns and cities were more accessible and social opportunities greater. They tended to be optimistic by nature, and to think that the good times would last forever, that the wool market had reached a new plateau. They believed this should prompt a change to the old ways of doing things, and adoption of a more expansive and less parsimonious management style. But the previous generation felt from bitter experience that prosperous times could be ephemeral. They were to be proved right. Certainly, advertisers of luxury goods and services were targeting the newly affluent. In 1951 York Motors were offering the 4.5-litre Bentley, pointing out: 'Now is the time to buy! More and more successful primary producers are choosing Bentleys. Today, fewer bags of wheat or bales of wool are required to purchase a Bentley than ever before.' Major and Madame Jacques HE Cable of Mt Dandenong in Victoria were seeking students for Le Colombier Swiss Finishing School, a 'private and select finishing school for the study of languages (French and German), continental culture, deportment and domestic science'. Peapes in Sydney were ready to kit out young woolgrowers for the many balls and dances – offering full evening dress-suits in black wool barathea and corded silk facings for £31 nine shillings, and patent leather dancing pumps for five pounds. Real estate agents were seeking buyers for a fully furnished two-bedroom cottage at Palm Beach for £6500.

Most of the 20-year-olds resisted the temptations of the high life, with the possible exception of cars, trips to the city and ownership of modestly performing racehorses for the local meetings. Some, to the strong disapproval of their fathers, indulged themselves in Jaguar XK sports cars, which they drove at excessive speeds up and down the newly expanded stretches of bitumen roads; many more bought massive V8 Ford

Customlines, Buicks, Chevrolets or Pontiacs. Sydney nightclubs were well patronised; at the top end of the market were Romanos and Princes, which were more like restaurants with a floorshow. Then came the Celebrity Club, Carl Thomas's and The Roosevelt at Kings Cross. When all else had closed, people retreated to dance the night away in the dark recesses of The Hayden in lower George Street, ruled over by the notorious Jack Spooner. His specialty was risqué songs with lyrics such as: 'The admiral's daughter / is down by the water / she's out to get your *dinghy*.'

Others dabbled in modest forms of diversification; a good few started small beef-cattle studs. To add some social tone to this activity, stock agents Dalgety and Co. secured the services of Lord Lovat, from Beaufort Castle in Inverness Shire, Scotland. They announced: 'Orders for the import of Beef Shorthorn and Aberdeen Angus cattle entrusted to Dalgety's will, unless otherwise instructed, be placed in the hands of Lord Lovat for his personal selection and report.'

Some Australian primary producers have always had a weakness for faddish new enterprises. In later years it was to be emus, alpacas and low-line cattle, but the popular diversification in the late 1950s was Landrace pigs. This was a European breed with exceptionally long bodies, and it was expected to transform the economies of pork production. Over the years it has been common experience that early entrants into fad enterprises reap rich rewards, and later entrants get burned. One early entrant was EJ Galbraith of Kyabram in Victoria, who purchased a Landrace boar and three sows from Ireland in 1957 for £1700, and within two-and-a-half years had grossed nearly £100,000 from their progeny.

Less fortunate was the grazier from west of Longreach, whose 22-year-old son lacked drive and initiative. In fact the young man was useless. Hoping to cure his lethargy, the father sent him off to the Brisbane Exhibition with 2000 guineas to buy a new Shorthorn bull. Sadly, there he fell into bad company, and he returned to the station two weeks later, hung-over and broke. 'When does our new bull arrive son?' 'Well Dad, there is no bull, but I did manage to buy this extraordinary pig.' Eventually the pig, and an account for 2500 guineas plus freight, arrived. The only pig the old station owner had ever seen was his breakfast bacon or a wild boar in the swamp, and he developed no great affection for the animal, which lay for months nuzzling its swill by the kitchen gate and eating scraps. It represented his family shame, his inability to produce a worthy heir, so each morning on his way to work he fetched it a smart kick in one of its many ribs. The pig eventually tired of this, and one morning jumped up without warning and drove the grazier through the kitchen wall, putting him in hospital for a week and causing another £500 worth of damage. Its life thereafter was short, but the meat proved inedible. The son himself was lucky to survive.

The wool boom of the early 1950s had some unfortunate social consequences. The tag 'grazier' was used excessively and indiscriminately by the press, and grew to have connotations of wealth, arrogance, autocratic behaviour and immoderate spending. Despite changed circumstances, these public perceptions were to persist right through until the crises of the 1970s, and proved socially and politically damaging. In addition, the merry sound of strongly flowing revenue streams attracted the unwelcome attentions of the taxman.

The 18th century English statesman Edmund Burke once said: 'to tax and to please, no more than to love and be wise, is not given to any man.' So it was not surprising that in 1950 woolgrowers were horrified when they learned that the Government intended to impose new taxes totalling more than 40 per cent of their gross revenue from wool. They would have been even more shocked had they known that the Country Party, led by Treasurer Arthur Fadden, which was supposed to be dedicated to their interests, was leading the tax charge. During 1950 a key government advisor, Sir Douglas Copeland, suggested to the Menzies–Fadden Government that a 33 per cent export tax should be placed on wool to help control inflation. This was to be on top of a 7.5 per cent special levy to fund a marketing plan designed to put a floor under the market. CW Russell, a tough and successful Queensland pastoralist and then Country Party member for the seat of Maranoa, recalled the circumstances: 'Fadden was concerned about the money they [woolgrowers] had and claimed they could not spend it properly, and that it would be largely wasted on riotous living. In the party room Fadden regaled us with instances of how the growers were squandering money. I entirely disagreed with him, because the vast majority of woolgrowers and primary producers are thrifty. All they would have done was to carry out much needed development on their properties. He felt it was necessary to skim off the profit of woolgrowers, and did so by a smart change in income tax laws.'[1]

Following an avalanche of protest around the bush the proposal was modified to a Wool Sales Deduction tax of 20 per cent, to be credited to the grower as a provisional tax payment against next year's income. It was the start of Provisional Tax, which remained in place for the next 50 years, and was to greatly inhibit the start-up of small businesses. The 7.5 per cent special levy was retained as a means of raising capital to operate a reserve price plan, which in the end was never implemented, and the money had to be handed back.

So, despite concessions, woolgrowers were still to pay tax of 27.5 per cent on their gross revenue from wool, in addition to income tax and other onerous taxes. Federal Land Tax was still in place, and rates were increased in the 1952 budget. GG Hawkes, President of the Stockowners' Association of South Australia, told of one farmer whose assessment, as a result of increased land values and rates, had risen from £307 to £3100. The most damaging of all taxes, in terms of fragmenting efficient production units, were state and federal death duties, which had to be found in cash soon after the testator's death. In New South Wales an estate worth more than £50,000 was charged duties of 24.5 per cent of capital value. Those over £100,000 were charged 39.1 per cent, and over £200,000 the rate was a crippling 46 per cent. The rapid increase in land prices following the lifting of controls in 1949 meant many more estates were falling into the higher brackets, and there was no escape for ailing or ageing graziers who gifted properties to their children; if they failed to live three years full duty was charged. Many of my contemporaries, born in the 1930s, had to sell all or part of their family property to pay duties. A good few were left with areas that were too small, and went under during the hard times in the 1990s. Indeed, federal and state governments bore a heavy responsibility for the crisis that was to envelop wool growing in that decade.

The simple expedient of defraying income tax by undertaking deductible property improvements was unavailable. Fencing material and steel were virtually unobtainable because of the effects of wartime restrictions and strikes on the coalfields, and the government refused to allow the import of wire from America. So it was not surprising that woolgrowers became paranoid about tax, and that avoidance became a popular art form. There was a widespread return to barter trade, and cash was king. In the early 1960s I was a professional sheep classer, and had to report to a grazier in the Bourke district on his sheep flock. He was away sinking tanks (in exchange for sheep, using plant he had bartered with a neighbour), and when I found his isolated camp he offered to pay my fee in cash, taking down a sugar bag that was hanging under the branch of a Wilga tree. It contained £3700 in notes. When I showed some surprise, his only comment was, 'Bugger that tax man'.

In the days before the pork industry aggregated into huge production units, many farms ran a few pigs round the house to utilise kitchen scraps and damaged grain. To cater for this trade, during the 1950s and 1960s the major livestock agents conducted regular pig sales in many small towns through south-western New South Wales. Settlement at these sales was strictly in hard cash; every vendor was a Smith or Jones, and any well-dressed stranger seen around town on sale day was given a wide berth under suspicion of being from the tax office. During the early 1960s I classed sheep for another grazier in the Hillston district, where I used to stay overnight so as to get an early start. Wandering around the farm buildings one morning, I was intrigued by a shed securely fastened with a rusting chain and padlock. Peering through cracks in the slab wall I could see nothing. As we got to know one another better he opened the doors to reveal a powder-blue imported Dodge sedan in the elaborate gull-wing style of the 1950s. It was up on blocks and had done no more than the delivery distance from Melbourne. The car had been accepted in payment for sale of wool 10 years earlier, never declared as income, and the owner was in such fear of being caught that he had locked it away from prying eyes. A few years earlier the wife of a station manager north-west of Bourke noticed her small children playing in the yard with pieces of paper. On closer inspection she found they were five and 10 pound notes. The station's owners had locked the office when they moved to Sydney in the early 1950s, but the curious children had prised open some floorboards and discovered two Globite suitcases full of cash that was being allowed to cool.

The prosperity that attended the early 1950s revitalised a social life that had been dormant for 10 years. Balls and dances – white tie and tails for the former, black tie and dinner suit for the latter – became frequent events in local halls or the large homesteads, where they danced the popular Brazilian Samba. Trips to Sydney for the June Sheep Show were crowded with social events. The Pickwick Club on the corner of Bligh and Hunter Streets was a popular venue. In 1952 it was the location for a sit-down dinner for the Corriedale Sheep Association, an informal supper for the New South Wales Sheep Breeders' Association, and a cocktail party given by rural merchandise supplier F.W.Williams and Company.

Wool Brokers Commonwealth Wool and A.M.L.&F. Co. gave elaborate parties for their clients, and it was back to the Pickwick Club on ram sales night for GBS Falkiner to host a 'large entertainment' for customers of his *Haddon Rig* Merino stud. Polo was revived, polocrosse clubs were formed, and picnic races regathered their momentum. Women dressed up in the 'new look' full skirts, and men wore the regulation uniform of the time – Donegal tweed sports coat, a tie, grey slacks, and Baxter elastic-sided riding boots. They sported short-back-and-sides haircuts cropped close up to ear level, always covered with a hat. Men of that and the previous generations would not even step outside the door without clamping a hat on. The old country barbers found it difficult to change their habits. When the vogue for long hair arrived in the 1960s, many a young man, who had asked for his hair to be left long back and sides, walked out of the barbers with the same severely cropped cut his father had worn.

Old-fashioned gentlemen's clubs were in their heyday. Establishments such as the Waradgery Club in Hay, the Murrumbidgee in Narrandera and the Riverine in Wagga Wagga were modelled on English clubs, right down to billiard rooms, leather furniture and the exclusion of women. Many had hefty joining fees and waiting lists for membership, applicants went to ballot, and it was not unusual for prominent citizens to be excluded by use of the blackball because their backgrounds were not right or because of some youthful indiscretion. In the 1970s the Waradgery Club made a substantial concession by allowing mixed company at dinner on Saturday nights. At one of these I wandered into the men-only bar and remarked to a crusty old member what a great idea it was. He sniffed, 'It's just like a bloody cafe.' Rowand Jameson, then manager of the local Merino stud *Uardry*, wrote of the Waradgery Club in the 1950s: 'At one stage of its history the club had been in financial trouble, the only revenue being from members subscriptions and bar profits; at a General Meeting it was decided that the only remedy was to drink their way out of debt. They must have made a good job of it, as the Club was financial when I joined in 1951.'[2]

Homesteads on the big stations were reawakening. Suzanne Falkiner wrote of life at her family's Merino stud: '*Haddon Rig* in this period resembled life in a small village. A small school was run in a cottage on the creek bank for the married station hands' children. There was a *Haddon Rig* horse racing track, a cricket ground and tennis courts, and the station fielded its own football, cricket and polo teams, and in the 1950s GBS Falkiner introduced the screening of feature films once a fortnight on outside screens.'[3] In Queensland Angus Baxter, assistant general manager of the Australian Estates Company, and formerly manager of *Cambridge Downs* at Richmond, where he succeeded my father, appeared before a Royal Commission on Pastoral Lands Settlement. Defending his company's stewardship of their pastoral leases, he told the commission that at the massive *Thylungra* station in the Quilpie district they had provided a school with a qualified teacher, a movie theatre, and were commencing a £50,000 plan to upgrade employee housing. They even intended to provide accommodation for married men. In north-western Queensland during 1951 the first sod was turned in an unusual community project. The Winton Shire Council was rebuilding the North Gregory Hotel, which had burned down in 1946, at a cost of £87,000. They had obtained government approval to run the hotel, and the substantial brick establishment was to be an oasis of comfort, coolness and cleanliness in an isolated district devoid of other amenities.

As living conditions were improving, so was the traditional stratification of bush society changing. Admission to polite society had been very much on the English model, depending upon family background, education, manners and ownership of land. This was being turned around in the newly affluent bush of the 1950s. There lived in the Nyngan district at that time a knockabout station worker called Billy Richards, a wiry, quick-moving man with a high-pitched voice that delivered a torrent of words in a staccato style. There was a hint of flashiness in the curl to the brim of his hat, and he tended to live off his wits. Billy undertook all sorts of contract work for my father at *Buckiinguy* – droving, horse breaking, lamb marking, cattle mustering. By dint of hard work, thrift, cunning, and a sharp eye for the margin in a deal, Billy had become a property owner by the 1950s, which changed his social status but not his social background. He also became a buyer of stock from *Buckiinguy*, which was important symbolically for both he and my father. The first visit with his new status posed an exquisite social dilemma for my mother, who was of the old school. Where to give him lunch? The station hand's mess was no longer appropriate for a property owner, but how would he handle the white damask and silver butter knives in the main dining room? She solved it neatly; lunch was taken informally, quite out of normal custom, at the big pine table in the gauzed breezeway.

This decade of change saw the passing of many of the old 'marksmen'. Born in the bush beyond the reach of compulsory education, they could neither read nor write, and often spent their whole lives working on one station. Precluded from advancement by their lack of education, many were superb stockmen and perceptive in reading country and seasons. Though sometimes taken advantage of, their intuitive skills were more often appreciated, and station managers, with whom they frequently developed a special relationship, treated them with great dignity. I remember one of these proud old men telling me in the 1960s: 'I was born here. I've walked over the place, ridden over it, driven a car over it, now I've been flown in a plane over it, and I'm happy to die here.'

Jim Roberts at *Buckiinguy* was one of these. Of great but uncertain age, even when walking became difficult he could still swing onto a horse with the agility of youth. With twinkling eyes, neat white moustache and rubicund cheeks, he was always immaculately turned out in jodhpurs, concertina leggings, sleeves buttoned to the wrist and a waistcoat, even in summer. My father thought him a bit of a 'Jonah' because, even at the height of the 1940–45 drought, he invariably carried a rolled oilskin on his saddle. He left the property only once a year to visit a son next door at Christmas, driving a massive canvas-hooded car of 1920s vintage that was faithfully cleaned and polished every weekend. Jim would not readily admit to his lack of education, so for the 30 years that they worked together he and my father played out a little charade. During weekly visits to the woolshed area where Jim lived, my father was invited in for a cup of tea, during which Jim would pretend to have lost his glasses and go through an elaborate ritual so he could have his mail read to him.

One of the best known employees of Australian Estates was 'Gudgeon', also known as Billy Cambridge, who was the offspring of a Pacific Islander brought to the canefields during the blackbirding era and had spent all his life on *Cambridge Downs*. Though uneducated, Gudgeon observed a strict formality of manner, and greeted each new manager with proper ceremony. No English director of the company would dare visit the property without calling on Gudgeon. Many years after Angus Baxter had left *Cambridge Downs* and become a businessman in Brisbane, Gudgeon heard he was getting married and sent a new quart pot and saddlebag as a present. Baxter treasured them, and they remained on his office wall for the rest of his career.

Prospective wool buyers inspect the clip prior to auction – circa 1950s *(Peppin Heritage Museum, Deniliquin, NSW)*.

Some of the old marksmen were highly intelligent, and had their own way with words. Rowand Jameson recalled Paddy at *Evora* in the Blackall district, who insisted on calling him 'Mr Jimmerson': 'Paddy could not read or write and depended on his "missus" in this respect, but he could sign his name, given plenty of time. One day I asked him the diameter of the fluming into an earth tank that had to be renewed. Paddy did not know but, not wanting to be beaten, said: "I don't know what she is in the diameter, Mr Jimmerson, but she is a fair way around the circumstance."' Paddy was no fool. Jameson recalled: 'We were discussing a young chap who was a very good worker but a bit unstable and always moving to another job. I remarked to Paddy that a rolling stone gathers no moss. Paddy thought for a time and then said: "That may be right Mr Jimmerson, but you have to remember that it's the old standing stump that's always pissed on".'[4]

As older characters were leaving the scene, new ones were entering. Though few of the post-war immigrants found their way to the inland stations, some from the Baltic countries joined fencing and rabbiting teams until they got themselves established. Some of these tough, reticent men with blond hair and hard blue eyes, claiming to be Latvian or

Estonian, may have had more dubious origins, but they were hard workers and were accepted readily into the station communities. One scorching November day in the 1950s I was riding past the rudimentary canvas home of a fencing team out the back of *Gowan*, part of the *Terrick Terrick* complex, when one of these Aryans emerged from the tent looking a little dishevelled and waving a bottle in his hand. He greeted me with, 'You come in unt haf cherry brandy – is good for you'. When I refused he extended the offer: 'Then you come in unt haf woman – we haf two good ones here.' Aged 18 and painfully shy, I rode on with a wave of thanks.

Of course there was nothing new about migrant labour in the bush. Pioneer pastoralist Sir Samuel McCaughey came from Ballymena in Ulster, and sponsored many young men from that province in the early years of the 20th century, and as late as the 1960s there were still newly arrived 'Ballymena Men' working at *Coonong*, a McCaughey family property near Narrandera. A story was told of McCaughey: 'Word got about that he was always ready to give a job to a man from his home county. One day a Chinese man applied for a position. "Me come from Ballymena, boss," he said. McCaughey was so tickled that he gave him the job.'[5]

Life was becoming easier on the stations as ever more efficient chemicals were available for the control of blowflies. In 1954 the chlorinated hydrocarbons Aldrin and Dieldrin appeared; these gave protection against flies for an incredible eight to 10 weeks. Then in 1956 came the organophosphate Diazinon, which extended the protection for up to 28 weeks and was still in use in the 1990s.

Life was getting better, and woolgrowers were enjoying themselves again. No sector of Australian society deserved it more, but sadly their respite was to be brief.

CHAPTER SIX

A DECADE OF CHANGE

Change your opinions,
keep to your principles;
change your leaves,
keep intact your roots.

VICTOR HUGO (1802–1885)

The euphoria of the 1950s soon evaporated when woolgrowers fell on seriously hard times during the following decade. There were powerful social and economic forces at work that, during the 1960s and 1970s, were to sweep away the station life that my father and I had been born into.

A number of things contributed to a fall in wool prices. New types of synthetic fibres were becoming readily available, their very newness was seducing consumers, and wool was losing its share of world apparel fibre markets. The opening up of huge areas of new land in South Australia and Western Australia, the rapid spread of improved pastures that could carry more sheep, and increased productivity following the demise of rabbits saw sheep numbers and wool production reach new highs. The world was overflowing with wool and the supply pipeline was choking.

To make matters worse, the industry's emerging problems had triggered an outbreak of wool politics, and internal bickering was becoming exposed in the city press. The almost reverential feelings that city people once had for the bush and its people were beginning to fade. The growing immigrant population, who were doing so much to broaden the nation's outlook, had fewer connections to the inland areas than earlier generations of Australians.

Hard-working woolgrowers found the downturn in their affairs difficult to cope with in emotional as well as financial terms. After all, they had done all the right things – expanded their flocks, improved their properties, boosted their sheep genetics, reduced the amount of labour they employed, denied themselves and their families many small luxuries and worked harder, all for less tangible reward. The most grievous blows were loss of self-esteem and the realisation that the icon status of wool and wool growing in Australian society was being challenged for the first time in its long history.

In the late 1960s I was newly divorced and, seeking to rebuild a social life, took every opportunity to visit Sydney. One Sunday afternoon I was at a party in Cross Street Double Bay, dressed in standard bush uniform of moleskin trousers, elastic-sided boots and blue cotton Oxford shirt. The other male guests were sporting the fashions of the times – shoulder-length hair, body shirts and flared trousers. Without any warning one of them stalked over and launched into a verbal assault: 'What are you doing swanning round the Eastern Suburbs drinking booze, when we are paying subsidies to all you bloody woolgrowers?' Of course I made robust reply, but it stung and to this day the memory still hurts. The ground had changed. The rules were different. The bushman in the city no longer had hero status, could no longer boast about what he was contributing to the nation.

In 1956–57 wool had contributed 50 per cent of Australia's foreign exchange earnings; by 1970 that had halved. Wool represented only four per cent of gross national product, and woolgrowers were only one per cent of the population. At this time the urban population contained a growing proportion of politically and socially aware university graduates, and there was a much wider interest in government policy and taxation issues. Once, those who boasted of avoiding tax were admired as smart fellows for outwitting the government, but it was slowly dawning on the first of the baby boomers that subsidies, tax-breaks for business and the like were at the expense of the general taxpayer. Tax and other concessions for woolgrowers had become numerous. They included a subsidy on superphosphate, generous provisions for writing off capital expenditure, debt reconstruction, farm adjustment schemes and drought assistance. There was also Commonwealth support for wool promotion and research, provision of working capital and loan guarantees for a wool reserve price scheme, and a subsidy to support a wool price-averaging plan. As prices plummeted and the industry sank into recession, wool was front-page news in the metropolitan press, and the tax implications of what was happening became visible. City folk were beginning to view woolgrowers as bludgers on the system rather than outback heroes. As *The Pastoral Review* said in a July 1970 editorial: 'The wool industry will have passed an ignominious milestone in this year of grace marked by tacit acceptance of a direct subsidy on production.'

It marked both the changing times and the closing of an era when the old Metropole Hotel, which had stood at the corner of Bent and Young Streets in Sydney since its opening by William Paul Small in 1890, shut its doors for the last time in 1970. Favoured stopping place for successive generations of pastoral families, thousands of consignments of wool and drafts of rams had been negotiated among its potted palms and worn leather chairs. The white-jacketed waiters in the downstairs lounge were infamous for their studied indifference, particularly to nervous young men attempting to buy their new girlfriend a Pimms. Many a romance had begun, and many were broken, in the precincts of 'The Pole'.

Society was changing quickly. It was as if people had spent the 15 post-war years trying to recreate pre-war conditions, then, having failed, their resistance to change crumbled away and new things flooded in. Oral contraceptive pills went on sale in the US during 1960. Yuri Gagarin flew in space. The Cuban missile crisis came and went. John Kennedy built his make-believe Camelot and was assassinated. The first US troops arrived in Vietnam and the protest movement was born. The Cultural Revolution convulsed China, while Beatlemania gripped the world. The Six-Day War was fought in the Middle East and student demonstrations paralysed France. Neil Armstrong walked on the moon as hippies donned beads and caftans and 400,000 people celebrated at the Woodstock Festival. The young took over the world.

At home on their farms and stations woolgrowers were largely cushioned against the impact of social change. They belonged to a conservative industry with conservative values, so change tended to be effected in a more measured way. But they were not immune, and the decade saw huge changes to their lifestyles, mostly resulting from practical

The advent of the motorcycle on sheep stations saw the demise of sheep work on horseback *(F.S.Falkiner archives)*.

things such as connection to mains electricity, access to television, all-weather roads, automatic telephones and increased mechanisation.

Home-generated electrical power, mostly 32- or 110-volt systems with banks of wet-cell storage batteries, was being replaced by 240-volt mains power. Bush people were now able to enjoy the simple conveniences city folk considered a normal part of life: instant hot water, washing machines, electric irons to replace the lethal petrol jobs, constant bright lighting, electric stoves that didn't require lengthy warming or hours of wood cutting, and record players. Later came television and, most important of all, air-conditioning.

In 1962 Euroa, Victoria, district woolgrower and shire councillor Max Brook joined with 14 other local farmers in persuading the electricity supplier to run a spur line to their properties. The cost was funded by a loan from the State Savings Bank, and each farmer was required to buy three major electrical appliances. Max and Irene Brook bought an Astor tablegram, a Kelvinator refrigerator, a washing machine, and a Sunbeam mixer for the kitchen.[1]

During 1960 and 1961 television was coming to the bush. In Victoria, stations opened at Bendigo, Traralgon and Shepparton. At first adoption was slow; there were only 26 sets in the Euroa district in 1960. The American Wild West rode into homesteads through shows such as *Rawhide*, *Bonanza*, *The Rifleman*, *Paladin* and *Gunsmoke*. At an only slightly more sophisticated level were shows such as *77 Sunset Strip*, *Route 66*, *Bandstand*, *Bobby Limb* and *Sunny Side Up*. The pull of a black-and-white screen was strong enough to empty station offices after dinner, where previously managers and overseers had toiled away for hours.

Sport was a real bonus. In 1960–61, when Ritchie Benaud and Frank Worral were resurrecting test cricket through a scintillating summer, I drove from Sydney to Moree, then a full day's trip, while the Adelaide test was in progress. Through the towns along the way – Singleton, Muswellbrook, Murrurundi, Gunnedah, Boggabri and Narrabri – cafes, service stations and shops lay open and unattended. Everyone had gathered round the few television sets available, which were confined to electrical store windows and hotel bars, to watch the cricket.

Formed all-weather roads, constructed with bitumen sealing or gravel, were thrusting out into areas where previously any fall of rain had left properties isolated. In 1961 Queensland Premier Frank Nicklin announced that road schemes were underway in the Peninsula, Gulf and Channel Country areas. Projects included bitumen surfacing of the Rockhampton–Longreach Highway, clearing and gravelling of the Cloncurry–Normanton road, and bitumen surfacing of the Bruce Highway from Brisbane to Cairns. In New South Wales the Newell Highway from Jerilderie to West Wyalong, notorious for its suspension-shattering potholes, was being sealed. These improvements had a profound effect on the quality of station life. Towns became more accessible; travel was less taxing for vehicles and people, who no longer arrived at their destination covered in dust. The trip from *Buckiinguy* to Nyngan, which once took us an hour and a half over rough and largely unmade tracks, could be accomplished in a comfortable 45 minutes. We had seldom visited town more frequently than once every six weeks; now trips became weekly events, and people could participate in town sport. The overhanging threat of being cut off from help at times of injury or serious illness was taken away. Mail deliveries became faster and more reliable, and there was no longer any need to hold emergency stores.

Motels began to appear in the early 1960s, and were an important addition to the comforts of travel. The old country hotels had many drawbacks. To be sure, there were some excellent places such as the New Albury, Tattersalls in Parkes, The Max in Moree, the Commercial, Castlereagh and Amaroo in Dubbo, the Royal in Bourke and Corones in Charleville, but many were dreary and depressing places. Rooms were dingy, drab, sparsely furnished, with floors covered with tattered linoleum. Unless you were at one of the better establishments like the Knickerbocker at Bathurst or Canoblas in Orange, which had central heating, they were bitterly cold in winter. In summer guests had the choice of sweltering in their rooms, suffocating under mosquito nets, or sleeping on the cooler verandahs and being carried away by insects. Many seasoned travellers adopted the practice of getting blind drunk every night before going to bed so that these trials went unnoticed. Communal bathrooms were usually crowded, a long walk from your room, and hot water frequently ran out. Telephones in rooms were rare. One of the greatest disadvantages of hotels for stock firm representatives, sheep classers, ram salesmen and others who drove long daily stages on business was that dining rooms were not open until 8 am for breakfast, so by the time they paid their account travellers could seldom get underway until 9 am.

Motel rooms on the other hand provided a bathroom, telephone, television, radio, a small refrigerator and the greatest blessing of all – an air-cooling unit. Privacy was also important; after a long day talking to customers, travellers could close the door on their own bit of space. Breakfast was served in rooms as early as 6.30 am so they could be on the road by 7 am. Early motels such as the one at Nyngan, first out on the western run, were soon filled with permanent bookings.

In New South Wales the society of towns and their surrounding districts was being transformed by the activities of licensed RSL, bowling and golf clubs that sprang up in the 1960s. Their blond brick architecture and purple-with-gold decor may not have been to everyone's taste, but the revenue they generated through poker machines was enormous. They used these streams of gold to give lavish support to local sport and to provide a standard of food and service, in air-conditioned comfort, not previously offered in the bush; all at a price few cafes or hotels could compete with. Membership was cheap and open to nearly all comers, so the clubs became egalitarian social hubs. Previously social drinking in towns tended to be segregated: graziers drank in one pub; shearers and bush workers in another; stock agents had their own favoured watering hole; as did public servants and bankers. The groups seldom mixed socially; these clubs changed all that, and brought society together.

Late in the decade the old manual telephone exchanges were being phased out and replaced with automatic systems. Smaller exchanges often worked limited hours, 8 am to 9 pm, and with only one operator were difficult to raise at busy times. Party lines, many of which had 12 or more subscribers each with a call sign (perhaps two long rings, or a long and a short), were notorious for listeners-in, whose presence was detected by soft clicks as they came on line; no conversation could be regarded as secure. Long distance calls were expensive, so were only used for urgent matters, and connections could be fragile. From western New South Wales it might have taken 30 minutes to be switched through to a Sydney number, and during conversation the line might drop out two or three times.

Subscribers and small exchange operators often developed interesting relationships. Those who were rude to the telephonist, or failed to give a Christmas present, might be constantly cut off or put on the end of queues. At *Boonoke* we were connected to the little Conargo exchange, from which the Rose family gave outstanding service. The lady who operated the tiny Mt Fairy exchange near Canberra for nearly a lifetime had Brigadier Broadbent of *Merigan* as a customer. She had known him when he was but a humble captain and persisted, much to his annoyance, in giving him that inferior title. They spent nearly 30 years correcting one another. When her exchange was eventually closed she retired a happy woman. For most of her career, placing a call to Sydney was a major event, but when the English film director Tony Richardson leased a local homestead for the making of his Ned Kelly film in and around the adjacent village of Bungendore, her world expanded. She took great pleasure in switching his frequent calls to London, Paris, New York and Rome, and announcing in her sweet sing-song voice, 'Hold the line please, Mt Fairy calling.'

A lot of station business was conducted through telegrams. The larger stations usually had a shortened telegraphic address; F.S.Falkiner was 'BOONRAM CONARGO'. Pastoral companies used private codes to protect confidentiality, and managers spent hours decoding lengthy telegrams regarding budgets and livestock sales. Incoming and outgoing telegrams were dictated over the phone, and hard copies of incoming messages, on distinctive yellow paper, were forwarded in the mail. All copies were retained, meticulously pasted into huge cloth-bound books. Telegrams were forwarded from local post offices in morse code – I can still see the stooped and shabby old telegraphist in Nyngan wearing an eyeshade as he tapped away at his key.

Woolgrowers' terms of trade declined right through the 1960s as the market deteriorated. At the same time inflation and an adversarial industrial environment in Australia were causing costs to balloon. Hours of work were difficult to regulate on stations because of the unpredictable needs of livestock, so traditionally there had been an atmosphere of give and take between employer and employee, with time off given in lieu of overtime. Then in 1962 the New South Wales Industrial Arbitration Act was amended to extend the 40-hour week to rural workers; other states followed, and Saturday morning work was abandoned. The fight to maintain slim profit margins forced woolgrowers to seek every avenue for shaving costs, and they concentrated on reducing the number of people employed by increasing efficiency and mechanisation.

Milo, near Adavale in south-west Queensland, was typical of the overstaffing that had developed in the 1950s. The station maintained one employee to the equivalent of 1400 sheep – one to 10,000 was to become the norm within 20 years. In 1952 *Milo* was 400,000 hectares on which ran 45,000 sheep and 2000 cattle, plus 'scrubbers' (unbranded adult cattle that were not on the books). It carried a staff of 45. There was manager Ken Nicholson, an overseer and head stockman. There were eight cooks catering for the homestead, station hands' mess, outstations and mustering camps. There were two jackaroos, a groom, houseboy, mechanic plus an offsider, two windmill experts, three men in the road-grading camp, five stockmen in the sheep mustering camp and five in the cattle camp, four boundary-riders living in huts around the run, plus a platoon of stockmen working from the homestead. A bookkeeper, storekeeper and governess capped this off.

Milo manager Nicholson was a tough boss, and it was reputed that 365 men, including shearers, went through the station books in one year. Hopeful job applicants arrived out on the mail run carrying their swags, and waited on the office verandah for a cursory interview with the manager and overseer. The successful would be told to dump their swags in the huts, the rejects were given a week's pay and put back on the mail lorry. In the virtually unregulated bush labour market of those and earlier times, managers held firm views about what manner of person was a suitable employee, which often had little to do with experience or performance. Old FS Falkiner once wrote to an employment agent regarding men he was seeking – a blacksmith, a couple of agricultural men, and a rabbiter. 'We don't want any aristocrats, broken-down swells and such like, but good men who will stop for some time and give satisfaction.' Around 1950 an advertisement appeared in Queensland rural papers seeking a station hand for a property north-west of Longreach. It concluded: 'Must be able to handle bad blacks. Only genuine enquiries considered. No bludgers, dagos, drongos, dingos, new Australians, synthetic yanks or Queen Street ringers need apply.'

Milo was near the village of Adavale on the Blackall-Quilpie road, one of the tiny sunburnt hamlets that serviced the big stations. An acquaintance of mine once spent a memorable hot summer's night at a similarly dilapidated pub in south-west Queensland during the late 1940s. Forced to stay overnight because of a broken axle on his International truck, he joined four others around the dinner table to eat some stringy goat curry. While serving the meal the perspiring and slatternly waitress, wearing nothing but a grubby petticoat and badly in need of some shaving, went around the table in turn. 'Now you're first tonight, you're second,' and so on. The innocent young city-raised jackaroo wondered what horrors were in store for him, and so was vastly relieved to discover she was nominating the order in which they were to use the limited bath water.

In pursuit of lower running costs the remaining station horses were the first to go as managers switched to motorcycles for mustering and droving stock. The most efficient and best-value motorcycles were made in Japan, and the switch was hastened by the gradual breakdown of post-war inhibitions about buying Japanese goods. Within a few years media stories about the bush and stations, which had always carried the image of a stockman on a horse, were being illustrated by a stockman on a motorcycle with one sheep dog across the fuel tank and another behind. The speed of travel by motorcycles meant the end of outstations and boundary riders. The bigger stations might have had three or four outstations from which one man cared for 8000 hectares and the stock running upon them. He patrolled his area each day, often in a horse and sulky, and phoned the head station every night to report and receive instructions for the following day. In early times their huts were rudimentary affairs of unlined tin with dirt floors. Boundary riders lived hard. The outstation hut at *Beaconsfield* in the Longreach area was typical: 'Walls six feet high of corrugated iron, a gabled roof generally, inside about six feet by eight feet, a deal box for a table, a bed made of boards, a couple of blue blankets, a tin jug and a basin, and a sixpenny looking-glass on a deal case were all the furniture.'[2] Cooking was done over an open fire, and the summer heat in these little huts could be indescribable.

In later years comfortable cottages were provided for married men, and many spent a lifetime and brought up families in these isolated locations. Some gained a sort of immortality by having the outstation and surrounding area named after them. One was George Rouse, who spent his career on the northern end of *Boonoke* up until the early 1970s. The area is still referred to as 'The Rouse's country'. Rouse was an Englishman who had migrated to Australian as a boy. During my time at *Boonoke* he was unsure of his age, but thought he was in his early 60s. The company's superannuation fund was anxious to know his true age, so sent off to England for his birth certificate. We had discussed retirement with Rouse, but he was adamant he loved his work and would stay at the outstation until he dropped. As he seemed to be in rude good health, full of enthusiasm and energy, we happily agreed. When the birth certificate arrived we found he was born in Dorset, the offspring of a farm labourer and a domestic servant, and was nearly 70. Within weeks of this discovery his health and strength seemed to evaporate. He took up a bored retirement in town, and died a few years later. In 1987 the remains of the little cottage, which had stood empty for years, were obliterated in a devastating bushfire. A few days later the daughters of the late George Rouse gathered to make a final pilgrimage and shed tears over the smoking remains of their childhood home. Stock and country were never better looked after than in the days of the outstations, for they were being constantly checked by experienced and dedicated men and any problems were picked up early.

During the 1960s the structure of station staff was changing as big teams of single men were being replaced by fewer and more expert married men. The old station hands' huts were closed down, and the change created a more settled social atmosphere on the stations. Gone were the boisterous ways and profanity of the old days when there was no opportunity for a station worker to marry, unless he was prepared to buy a house in town and leave the family there while he spent his week out on the station. One employee described the arid weekends for single men at *Beaconsfield* during the 1920s: 'The one day off each week was a dreary, boring day which cast a gloom around the homestead. Anderson [the manager], who was never happy except when he was working, preferred to spend the day branding colts or perhaps killing a bullock. Most of the men were inclined to help, after doing their laundry work, washing their moleskins.'[3] Fred Knight told of one of the old types at *Boonoke*: 'Jim Chaffey chewed plug tobacco and was able to eject it in a long, straight line out of his mouth, through a gap in his teeth. He was a great trainer of sheep dogs, and once took one of them to the Royal Melbourne show. At the trials, his language was such that the judges intervened, but Chaffey said that if he did not swear, the dog would not understand him, so the dog carried on without him and was awarded a special prize.'[4]

The advent of mains electricity put an end to the great unwashed. It was no longer necessary for workers to cut kindling and light up a chip heater, or stoke up a coke donkey, to bathe when they came home at the end of a long and tiring day. Prior to that there was an excuse, and cleanliness was optional. Though they were a minority, some station workers avoided washing themselves or their clothes for long periods. Cuthbert Featherstonehaugh, manager of *Brookong* in the 1890s, told of a dingo trapper called Dick Lynch: 'Just when I had one paddock ready to take sheep, five wild dogs were seen on it. We had a splendid chap named Dick Lynch who used to get the dogs wholesale. We declared the smell of old Dick going round the boundary kept the dogs away – he only changed his clothes once a year, and that was the only time he ever washed.' At *Buckiinguy* in the 1940s there was a station hand called 'Scotty' Bates (in fact he was a Welshman) who studiously avoided bathing. Every store day his workmates would buy him a bar of Sunlight Soap as a present, but he never took the hint.

Station bookkeepers were another casualty of the 1960s and the improved communications. They usually kept a cashbook, journal and ledger, paid wages, settled suppliers' accounts, attended to correspondence, and sent monthly ledger balances off to head office. They had charge of the ordering and issuing of stores, station supplies and fuel. Station tools were usually kept in the store and booked out each day. Bookkeepers saw their role not so much in preparing accounts but in reducing costs, and their parsimony was legendary. If the cook sent for 20 pounds of flour, they would issue 10. One famous bookkeeper later became accountant for a firm of city motor dealers, and was fuelling up a car for a new owner. When told it would take 20 gallons he responded: 'Wouldn't 10 gallons do you?' It was commonly said of bookkeepers that 'he acts as if he owns the bottom wire', meaning he acted as if he was part owner of the business – owning the bottom wire of every fence. We were often reminded of the scene from an early Dad and Dave movie, in which Bert Bailey as Dad was resisting son Dave's arguments for buying a new reaper and binder. As he turned to walk away he shouted: 'You can't have a new reaper and binder, and that's final. You'll have to fix the old one with wire,' then as an afterthought, 'and make sure you don't use that new wire either.'

A bookkeeper had few prospects for advancement; it was usually a dead-end job and a solitary life, for there was no accommodation for married men and little society other than the station society. So a strange assortment of people took up these posts. Some had failed in commercial life; others were escaping cities for a variety of reasons or carried some kind of handicap. There were those with interesting backgrounds; I remember one quiet old chap who was a former British naval officer, and as a teenaged midshipman he had piloted one of the boats onto the beach at the Gallipoli landing. A few were unable to cope with the outside world. Some, like my father near the end of his life, were retired station managers who couldn't stand living in the city. Understandably, many were frustrated in life and went 'sour' on the job, snapping and snarling at everyone they came in contact with. There were alcoholics – men who were efficient at any job until they started on a bender. A former tax accountant who worked at *Boonoke North* kept a bottle of gin suspended on a string in the cistern of the office lavatory, and would spend the afternoon answering frequent calls of nature, nodding off to sleep, and answering imagined telephone calls. 'Backblocker', who wrote a column of advice in *The Pastoral Review* during the 1950s, warned: 'At one time the back country was looked upon as a sanctuary or inebriate asylum for men who could not hold a job in the city, hence a manager has to be always on the lookout for grog troubles.'

Fred Knight, who was a Melbourne solicitor, told of Mr Drew, formerly bookkeeper on the Falkiner station *Moonbria*: 'Twenty years later he arrived in Selbourne Chambers in a filthy suit, so dilapidated that the canvas of his coat was showing. He demanded help because he wanted to get into an Old Colonists' Home. I gave him an old, but respectable suit, half a crown, and procured a caretaker's job for him. His new employer was so sorry for him that he gave him five pounds. He had a glorious spree on brandy, left the job, pawned my suit, and returned to me more dilapidated than ever. He was typical of the more educated men who spent their lives working as bookkeepers on country stations, where they could not get a drink. Many of them spent their holidays in good city hotels until their cheques had been cut out.'[5]

Up until the 1960s the F.S.Falkiner group of properties, like many others, employed a bookkeeper on each of their six properties, plus an accountant and secretary at head office. As the profitability of wool growing declined during the decade such infrastructure became unsupportable. Secure automatic phones and faster and more reliable mail deliveries meant that stations could operate using a simple cashbook and cheque voucher system, and were able to centralise the accounting function. Better access to towns and the switch from single to married employees enabled the prompt supply of stores from the newly opened supermarkets. The old station stores were closed down. Bookkeepers were no longer needed, and a colourful era of station life closed.

The 1960s also saw the symbolic passing of a traditional bush artefact – the quart pot. This was a small tin billy for making tea, oval or round in shape, enclosed in a neat leather case for strapping to the side of a saddle; a smaller metal cup, the pint, fitted snugly into the top. The three most important items in a bushman's gear were his swag, saddlebag and quart pot. In 1960 stainless steel Stanley vacuum flasks appeared on the market for £10. Previously the inside liners of vacuum flasks were made from glass, so would not withstand rough treatment, but the Stanley liners were stainless steel with a carbon steel outer casing. They were unbreakable and lasted a lifetime, so there was no longer the need to build fires during lunch camp to boil the mustering team's quart pots. Bush life was becoming more convenient.

CHAPTER SEVEN

CATTLE BOOM ... AND BUST

Tis summer time on Bredon,
And now the farmers swear:
The cattle rise and listen
In valleys far and near,
And blush at what they hear.

HUGH KINGSMILL (1889–1949)

During the 1970s the beef cattle industry in Australia enjoyed an unprecedented boom followed by an almost overnight bust of catastrophic proportions. Late in 1973, at the F.S.Falkiner properties in the Riverina, we purchased 800 splendid Shorthorn bullocks at $200 a head from one of the Dalgety's properties. They were magnificent beasts, lovely rich reds and roans in colour, glowing with good health; we were proud to have them striding around our paddocks packing on additional weight, and fully expected to make a handsome trading profit on their resale to the butchers. We ended up selling them at $80, and with the market plummeting were lucky to get out of them at that.

When wool prices slip into a protracted trough, as they did during the 1960s, woolgrowers often switch into cattle without giving sufficient thought to the long-term impact of their decision. In 1970, with wool prices languishing and their flocks unprofitable, woolgrowers joined an unseemly rush to buy cattle of any description, irrespective of price, quality, or the ability of their properties and pastures to carry large stock. Bank managers, wool brokers and farm consultants were urging people to buy, and buy they did. With finance readily available, stock agents acted quickly and were busy filling cars with potential buyers to scour the country for cattle. Stock routes and sheep paddocks soon filled with blacks and roans and brindles and dairy crosses and tropical breeds and everything in between, so long as it had four legs and produced beef. 'More mongrels than in the Sydney Dog's Home,' was the dismissive comment from traditional cattlemen.

The lift in cattle numbers was spectacular. In 1963 there had been 18.3 million head in Australia; by 1976 they had increased by 83 per cent to 33.4 million. In 1972 Australian beef production stood at 1.1 million tonnes, and was about to rise to nearly two million. Australians were chewing their way through an average of nearly 40 kilograms of beef each year, and unless appetites increased we were not going to get through the additional supply. So the volume available for export was about to double. Fortunately the doors to export markets swung open freely; demand was strong from the European Economic Community [EEC], the previously untapped but beef-hungry Japanese market seemed desperate for more 'Aussie Beef', and the US became our largest and most profitable customer. United States retailers and consumers were objecting to high prices for prime cuts, so President Richard Nixon got in on the act, temporarily easing the tariff on imported beef. The USSR even joined the rush, taking a trial shipment of 4500 tonnes of Australian product.

A dangerous atmosphere of invincibility surrounded Australian beef producers, and everyone thought the good times would go on forever. We forgot that the US and EEC both had big domestic herds of their own. When bullock prices at Newmarket in Melbourne reached a record $438 in 1973 there were calls for a lid to be put on prices, but the reaction from exuberant cattlemen was: 'The market isn't inflated – our cattle are finally making realistic prices.'[1] It was the standard reaction of Australian primary producers to price booms. Australia's Trade Minister Doug Anthony declared that rising living standards in Asia would continue to fuel record demand for beef, and a campaign began for Australia to build its cattle herd to 40 million over five years, and 50 million over 10 years.

Some voices were preaching caution. Regional politicians were developing concerns about the shift away from sheep, which were labour-intensive to run and provided more employment in their electorates, particularly in the shearing industry. The Bureau of Agricultural Economics issued a sonorous warning in 1972, which few either heeded or heard. Beef producers were unwittingly exposing themselves to a high degree of sovereign risk – because access to too many export markets was regulated through government-to-government agreements. The US had become Australia's biggest single market, taking 62 per cent of total exports and 31 per cent of total production, but their market was regulated through quota restraint agreements with the governments of supplying countries, and their legislation tied imports to levels of domestic production. Then in 1973 the OPEC oil-producing countries began to commit mischief by restricting oil supplies in a bid to lift prices, triggering severe economic problems in the developed world where lay most of the markets for Australian beef. In 1974 doors began to slam shut one after the other as the governments of beef-importing countries moved to protect their domestic industries. When finally the US closed its doors the results for Australian producers were devastating.

A good litmus test of the state of the Australian cattle market was demand for steer weaners bred in the beautiful mountain country around Benambra, Omeo and Ensay in Victoria. These well-bred young cattle were popular among producers further inland, who bought them for growing-out and eventual resale. In 1974 they averaged $135 per head. Then in 1975 the average tumbled to $51, and was to languish around the $40 mark for the following three years.

With deteriorating seasonal conditions across most of eastern Australia, some types of cattle became unsaleable at any price. Under a government emergency relief package, producers were offered two dollars a head to shoot surplus animals and bury them in mass graves. To this day many have never forgotten this distasteful experience, the blood, the horror, the waste of it all. Advertising campaigns were launched in a vain effort to stimulate domestic consumption, and the cooking pages of newspapers and magazines were full of recipes for various beef dishes. Australia's cattle herd began to reduce sharply, and within seven years had adjusted back from its peak of 33.4 million head in 1976 to a more sustainable 22.5 million. By the time of the cattle crash in 1974–75 the fortunes of woolgrowers had turned around, and once again they could hold their heads up and stand a round of drinks in the Royal Mail hotels around the bush. Repentant cattlemen were no longer referring to sheep as 'ground lice'.

Sheepmen who had switched to cattle had ignored some quite fundamental issues, and by doing so had learned a bitter lesson. Meat markets, particularly beef, tended to be more volatile than wool markets; the booms and busts were more pronounced – not least because the world trade was more regulated than the trade in greasy wool, and hence more susceptible to political interference. Some of the pastoral country – particularly those areas that grow perennial native shrubs such as saltbush – is unsuited to stocking with large numbers of cattle, which are more damaging to the natural environment than sheep. Many sheep properties had inadequate infrastructure to handle cattle, which are hard on watering points and fences.

Cattle are also more difficult to handle in drought than sheep, which are better foragers and continue producing wool and hold body condition on low-quality pastures. During the 1970s drought I watched a noted Gippsland cattleman standing among his emaciated stock, and heard him cry out in despair: 'Look at the bastards! All they do is stand around and bellow and shit, and then they die.'

The cattle crash of 1975 virtually put an end to the activities of a fascinating sector of the pastoral industry – the big cattle dealers. These men bought and sold huge mobs of cattle, often for small profit margins. Few of them owned large stretches of land, so when moving their stock to

Cattle penned for auction at Newmarket, Melbourne, in the 1970s *(National Library of Australia).*

market they relied upon an elaborate chess game using stock routes, agistment and leased country. A few stayed in the game, but a combination of low prices, wider use of road transport and better information on markets ensured they never again operated on such a large scale. These men lived at the very edge of risk, and traded on a vast scale right across eastern Australia. They might simultaneously be buying store cattle at Longreach and selling fat cattle in Melbourne. They might have thousands of head in play, or be trading so fast that they lose track of the numbers. They were always chasing grass, and were quick to move stock to wherever rain had fallen. They maintained active intelligence networks, which tipped them off as to where they might find cheap cattle or abundant grass.

The dealers included Jack Smythe and Keith Buffier, operating from Tamworth; Jack Ward from Harden; the Weedner family from Albury, and George Willox from Swan Hill. They were shrewd judges of trading margins, the weight of a beast, and people. They were not the shifty-eyed horse traders that many expected – for men like 'Gentleman Jack' Smythe their word was their bond, and they clinched weighty deals on a handshake. When I was heading up the F.S.Falkiner operation at *Boonoke* we bought steers from Smythe on several occasions – each time he told us how much profit he had taken on the deal, and how much he reckoned we would make. He was usually right. Livestock agents loved to have a dealer on their books as a client because of the number of trades they made. An agent would collect five per cent on the buy end of a deal, another five per cent at the sell end, and provide working capital at commercial interest rates. Often they would put a car and a driver exclusively at the dealer's disposal. Cattle dealers had a high tolerance to risk, and provided a useful function as buyers of last resort – when pastures had cut out, drought was looming and markets closed up, a producer could usually find a dealer who was prepared to buy stock at a price. They had nerves of steel. Jack Smythe had a thousand steers agisted at *Boonoke* at the time of the crash, and I phoned him to see how he was holding up. 'I've done a million dollars on paper, son,' was his reply. This was a considerable sum; when I asked what he was doing about it he responded, 'I'm still trading'. At the height of the crisis Jack Ward was reputed to have made the wry statement: 'I have always dreamed of being at a cattle sale where I was the only person who turned up. Well, now I have been – the only problem was that I owned the cattle.'

CHAPTER EIGHT

STATION TUCKER

On Monday we've mutton, with damper and tea;
On Tuesday, tea, damper and mutton,
Such dishes I'm certain all men must agree
Are fit for peer, peasant, or glutton.
On Wednesday we've damper, with mutton and tea;
On Thursday tea, mutton and damper,
On Friday we've mutton, tea, damper, while we
With our flocks over hill and dale scamper.

FRANCIS LANCELOTT, *AUSTRALIA AS IT IS*, 1852

Food was a constant preoccupation and topic of discussion on the old stations. In that respect they were not unlike other close-knit and closed communities, such as boarding schools and the military. Indeed, station people took tucker very seriously, though probably more in terms of its quantity than its quality. Nothing bred discontent more quickly than poor or scanty food. Up until the 1960s most people on stations had rations thrown in as part of their remuneration, hence the popular saying: 'The more you eat, the more you earn.' What TA Browne ('Rolph Boldrewood') wrote of shearers was true of most pastoral workers: 'They worked on the principle that the more they ate the stronger they would be. Digestion was left to take its chance.' There was also a tradition of bolting food down as fast as possible, because it was widely believed that someone who ate slowly also worked slowly. Shearers again led the way – as soon as their last sheep was on its way down the chute following the lunchtime bell, they were off lickety-split to the dining room. There they shovelled down two courses, probably roast meat with vegetables and gravy followed by a pudding, and raced back to the shearing board to smoke and doze until the start-up bell rang.

In early times station stores opened for one day a fortnight or a month, and issued rations of tea, sugar, salt and flour on a set scale. One dipper for each adult, half for each child. The old pewter or tin dippers that were used can still be found on station rubbish tips. In the middle of the 19th century the official weekly ration per adult was sugar (very black) two pounds, green tea (seconds) a quarter pound, flour 10 pounds, salt as required (allowing for the need to salt meat before refrigeration was available). Meat, killed on the station, was plentiful, though usually kept to a sheep per week for each family. There is a legend that Otway Falkiner was once called as a character witness for an unfortunate boundary rider caught by the local vicar committing an unspeakable act on a sheep. When questioned by the prosecution, Otway went into a long diatribe about how good the *Boonoke* sheep were. Drawn back on to the topic in question by counsel he was reputed to have said, 'Oh that – well, every employee gets a sheep per week, and we don't care whether they eat them or screw them.'

Things like jam, tinned meat, dried fruit, and condensed or powdered milk were outside the ration scale and had to be purchased from the station store. Opening times were strictly scheduled so if a family had guests, or their rations spoiled, then they ate meat and drank water until the next store day. Right through until the late 1950s stores on the larger stations were stocked like a general store in a fairly substantial town. Staples of flour, tea and sugar were bought in bulk by the companies and held in great tin-lined bins. There would be rows of tinned meats to vary the monotony of mutton – tongues, stringy bully beef, and the mysterious camp pie. There were four-gallon tins of bulk jam – usually plum, strawberry, or melon and lemon. When I was a child melon and lemon jam was cheap, and we were served so much of it that to this day the mere thought of eating it fills me with horror. My father had the same aversion to apricot jam. There were dried and tinned fruits, tobacco and matches, various herbs and spices, saddlecloths, stirrup leathers, harnesses, oatmeal, condensed and powdered milk, soaps, soda and candles. Cartridges, bullets, strychnine, cyanide and arsenic were locked in a safe. There was gum-arabic, saltpetre, raddle, axes, Scrubb's ammonia, large quantities of tomato and Worcester sauce, pickles, chutney and curry powder, junket tablets and jelly crystals, pills and potions for people and horses. I remember the cavernous interior of the station store as a wondrous place for a child. Shrouded, mysterious, delicious, dangerous, and redolent with what I imagined to be the scents of Asia.

Mutton and bread was the staple diet. Invariably a well-roasted leg or shoulder for dinner at night, with baked potatoes and pumpkin, with the remains eaten cold for lunch the following day. Grilled chops for breakfast, with or without onion gravy, with such monotonous regularity that they were called 'Holy Meat' – to echo the dismal cry when walking into breakfast: 'Jesus Christ! Chops yesterday, chops today, chops tomorrow and chops for evermore.' Sometimes there might be the variant of lamb's fry (and gravy) or rissoles (and gravy). Eggs were only available if one of the station wives was prepared to run chooks and the station was prepared to buy the eggs. Breakfast at the main homestead, as distinct from the jackaroos' barracks and station hands' mess, could be more elaborate. At *Coree* in the 1930s 'a number of hot dishes were laid out on silver entrée dishes: kidneys and bacon, mutton chops, crumbed brains, eggs, kedgeree or salmon fishcakes. In winter we had porridge.'[1]

Of course, there was also boiled mutton, corned mutton, mutton stew, shepherd's pie, mutton curry, mutton fritters or corned flap. Shanks were considered only suitable for the dogs. Station mutton, particularly in drought years, was low in fat content, dry, black, tough and stringy. To provide some variety, meat was often smothered in tomato sauce or pickles, but because they were costly, stations would ration supplies, much to the indignation of employees. Many of the stud sheep properties ate surplus rams, whose meat had a gamey tang about it. But you grew accustomed to the strong flavour. The Allen family, who for two generations worked for the Falkiners at *Boonoke North*, and had eaten no other meat for most of their lives, couldn't stand the meat when they travelled away from home – it was too bland for them. Mutton from wethers run on saltbush pastures was a treat; the densely textured meat was full of flavour.

All outside staff were required to take a sandwich lunch each day, even if they were to be working near the homestead. For nearly a lifetime my father ate three sandwiches, two of cold roast mutton and tomato sauce and one of jam, and a slice of Brownie. Brownies were vast fruitcakes with a moist, tawny texture, made without milk or eggs, and were standard fare on the stations. Lunch might be varied with tinned bully beef or camp pie, but they could be unattractive in hot weather when their fat melted into a greasy liquid. In any event, returned men had seen quite enough tinned meat during the war.

In earlier times bread was baked on the stations, and most old homestead kitchens had deep brick bread ovens alongside the wood stoves. Quality was variable – many cooks claimed to be able to bake bread, but few had the skills. Later bread would come out weekly on the mail truck from local town bakeries, sewn into hessian potato sacks, and always in the form of high-tin loaves. Once in the late 1940s my father failed to award the general station carrying to our mailman, so he retaliated by always stacking heavy goods on top of our bread, so loaves would arrive completely flattened. When roads were impassable and bread failed to get through, damper was relied on. A simple round loaf, kneaded from plain flour, baking soda and salt, damper featured as part of a bush diet from the days of early settlement. Everyone fancied themselves as top damper cooks, but their products could range from a nice cake-like texture to slimy, glutinous, inedible clods.

Shearers' cook, Peter Semmelhack, prepares the midday meal, 1955 *(photo by Bill Brindle, National Library of Australia)*.

Stations located in winter rainfall regions usually ran a small herd of milking cows to produce fresh milk, cream and butter, but in northern subtropical areas, where native pastures were low in protein, a milking herd could only be supported if the station was prepared to buy supplementary feed. In these areas it was Sunshine powdered milk or Nestles sweetened condensed milk in tins, which took some getting used to.

Once the Chinese gardeners had gone, diet on the isolated stations contained few fresh vegetables, no salad greens, and little fruit. The inevitable side effect was the affliction known as 'Barcoo rot', which was a form of scurvy. The first symptoms appeared when little cuts and scratches refused to heal, and finally a series of blisters would form on the backs of the hands. One old Queensland station man described the unpleasant course of the disorder: 'That original sore doesn't fester. "Fester" is too light a word to describe the throbbing agony of the burning sear. It's full of dead pus but you can't squeeze any out of it. It spreads. It's volcanic in shape, and little baby volcanoes form in a ring around the original eruption.'[2] Short of access to fresh fruit and vegetables, the treatments were various, but few worked. Epsom salts and 'painkiller' (which contained opium) were widely used in the 1890s. Barcoo rot defied Zambuc ointment, iodine, soap-and-sugar, and all the patent medicines. Some swore by an ointment made from a 20 per cent mixture of white precipitate used with any suitable base. But the best ready-at-hand cure was beer, taken internally in liberal quantities, which seemed to replace the dietary deficiencies. So it became possible for station workers to obtain a doctor's prescription for beer, which was necessary as liquor was banned in most of the station barracks. The monotonous diet also induced cravings for beef, eggs and chocolate. In my time jackaroos who hit town for the first time in months headed straight for the cafe rather than the pub, where we might polish off two T-bone steaks with four eggs, followed by ice-cream topped with chocolate sauce, with a quarter-pound block of chocolate to finish.

Station cooks were a race apart. Few were what you might call 'normal'. They ranged from the mildly eccentric through to hopeless drunks, and right on to raving lunatics. Competent cooks were seldom without some personality defect, and those with equable temperaments could seldom cook. It was unsurprising that most of the old-time cooks were to some degree 'cranky': their wage was only marginally higher than that of a general hand, the hours were long, kitchens were hellish infernos during summer, and the customers were always complaining.

Many were seeking refuge. There were prisoners on parole, defrocked priests and disbarred lawyers. Some were educated men with mysterious backgrounds – as a child at *Buckiinguy* I remember one old chap who moved among his pots and pans mumbling great slabs from Shakespeare and Milton.

Others were dirty and slovenly; the type immortalised in *The Shearer's Lament*:

The greasy cook with his sore-eyed look,
All covered with scabs and rashes,
Stuffed our holes with his half-baked rolls
And poisoned our dogs with his hashes.

Joan Austin Palmer told of one such character: 'These rock cakes had a sinister reputation as Bill, the hut cook, always worked surrounded by a cloud of flies, and any that were old or infirm became incorporated in the mixture.'[3] In the middle 1950s at *Wanganella Estate*, near Deniliquin, a general hand named 'Bunner' Henderson was roped in on short notice to cook for we three jackaroos. Bunner was not a strong believer in bathing or washing his clothes – he wore shirt and trousers week about, letting the spare set air on the clothesline. After a couple of weeks the grease-encrusted trousers could be observed swinging gently on the line, moulded into the shape of his legs. He was more meticulous in regard to socks and underclothes, which he laundered in the cooking pots. He also had an unpleasant habit of shaving in the washing-up water, with the result that plates often carried a film of grey whiskers. Bunner's menu was plain to the extreme, and eventually he indeed succeeded in poisoning us with one of his hashes. But, like many of his kind, Bunner was absolutely convinced he was an excellent cook. Once he had worked for a gang of Italian builders, and recalled with pride: 'Them eyeties loved my stews; they couldn't get enough of them.' The poor devils had little choice, because he served nothing else for three meals a day.

In the 1980s the barracks cook at *Boonoke* was Jack Collier, an engaging man and quite a competent bush cook. He had a habit of leaning against the kitchen door and smoking the butt of a cigarette to contemplate the jackaroos polishing off their bacon and eggs. One morning a surly big jackaroo came in, sat down, looked at the plate in front of him, and with a sneer of disgust pushed it away. Jack shot him a baleful look: 'That's all right about you young feller, you only have to eat the f … stuff, but I have to cook the muck.'

Strangely, many of the old bush workers I knew were deeply suspicious of good cooking. They had the irrational belief that it would erode their constitution – 'Ruin your gut' was the term – and an anonymous poet captured the attitude:

The song I'm going to sing to you will not detain you long,
It's all about a station cook we had at old Pinyong.
His pastry was so beautiful, his cooking was so fine,
It gave all a stomach-ache right through the shearing time.

But oh dear! I feel so queer, I don't know what to do,
The thought of leaving Fowler's Bay just breaks my heart in two,
But if ever I catch that slushy I'll make him rue the day,
That he ruined my constitution while shearing at Fowler's Bay.

There was a famous shearer's cook around the Riverina in the 1940s and 1950s who cooked in resort hotels during the off-season. Not only did he present superbly cooked roasts and puddings, he turned out fancy pastries, pies, cream-puffs, tarts, scones, cakes and puftaloons that were the envy of all. To top it off he was a jolly man with a pleasant personality who never became ruffled, and always wore a smile – a real gem. Yet when he joined a big shearing team there were always three or four old stagers who pulled out to join another gang, muttering that the food was too rich and would prove the ruination of them. Shearers were considered to be low in the social pecking order by women of my mother's generation, and when she corrected my table manners it was with the accusation that 'you eat like a shearer'. So I was much surprised when my father took me to the shed, where we lunched with the team, to discover they ate in the same manner as everyone else. I expected them to be tearing at the food with their hands and throwing the scraps onto the floor.

Some of the old bush cooks had a weakness for drink. As liquor was banned on most of the stations, after a few months they would get dry and cranky. 'Their skin would begin to crack,' as the saying went, and when this happened a wise manager would let them go so they could have a good bender and get it out of their system. One of these was Jack, who circulated around three stations in the Narrandera–Jerilderie area: *Boonoke North*, *Yamma* and *Widgiewa*. When on the straight and narrow Jack was such a pleasant chap, and his cooking so good, that he was always welcomed back when his horrors had subsided. Jack's most valuable possession was his false teeth, so when he felt a bender coming on he carefully parcelled them up and posted them off, addressed to himself care of one of the other stations on his regular circuit. You often had early warning that Jack would shortly be along looking for a job when the familiar parcel arrived marked 'Please Hold'. Sometimes Jack and his teeth moved in different directions; when they caught up with one another his current employer would witness a wondrous physical transformation. But tragedy was always lurking close at hand in the bush – during one Christmas heatwave Jack, deep in the horrors, mistook his direction and got off the train from Jerilderie at the then unmanned Widgiewa siding, while his teeth were waiting patiently at *Yamma*. To get away from the blazing heat he crawled under the railway platform, where he remained without food or water for some days until discovered by one of the mailmen, who smelled him. It was a very narrow escape.

Some found ways to circumvent the ban on liquor. Lemon essence was a favourite tipple, and bookkeepers were always on their guard against any excessive ordering. Many stations had fewer moral scruples about gambling, so allowed employees to use the phone in the homestead on weekends to place bets with the local illegal bookmaker, who usually operated from the pub. At *Terrick Terrick* in my time calls were made in the hearing of the manager or his wife, so one thirsty chap developed a code to smuggle grog out on the mail truck. He would tell the bookmaker: 'Put 30 shillings on the black horse on the last race.' Thirty shillings was the price of a bottle of spirits, the black horse confirmed that rum was required, and the mail truck left town at the same time as the last race.

I saw everything in relation to station food change very quickly from around 1960. Most of the eccentric wanderers and misfits vanished because the new society, and new economic realities, could not or would not tolerate them. Dietary knowledge became widespread. Wages for cooks improved. Respectable young women, who took a close interest in the quality and variety of the food they served, took jobs around the stations to accumulate a nest egg for overseas travel. The growth of supermarkets, better distribution systems, bitumen roads and faster transport ensured that fresh vegetables, fruit and a wider variety of foods became available. Access to mains power brought with it freezer units that could store meat and perishables for long periods. Gradually the old station stores closed and employees' wages shifted to the higher 'unfound' rates. Station people no longer relied on mutton and bread as their staple diet, no longer suffered from Barcoo rot, and lived healthier lives.

But lives just a tad more boring.

CHAPTER NINE

FUELLING CHANGE

The spirit of the time shall teach me speed.

SHAKESPEARE, *KING JOHN*

Most of the big stations were relatively isolated, though it was not the done thing to remark upon it. No matter how far from town and its conveniences you lived, there was always someone further out, and when things got tough it was common to say 'the going is hard here, but I don't know how the poor buggers in the back country are coping.' Yet isolation was a real factor that impinged on nearly every aspect of our daily lives. It increased the risk arising from sickness or injury, added to the cost of things, dictated what food we ate and restricted our social life. The big stations became self-sufficient, and the 'do it yourself' attitude added to the sense of living in a closed community almost monastic in nature. So in my lifetime, and my father's, any improvements in communication or transport brought about profound change.

In August 1923 CJ Cameron, manager of *Barratta*, Deniliquin, made a trip to class sheep at *Charlton* near Brewarrina in north-west New South Wales. He departed from *Barratta* early one Friday morning, took the steam train from Deniliquin to arrive in Melbourne Saturday morning, caught the express to Sydney that night to arrive Sunday morning, boarded the *Western Mail* on Monday morning, changed at Byrock for Brewarrina, and arrived grimy and exhausted at *Charlton* Tuesday night. Nearly four days for a trip that now can be done by car in one day.

But things were changing fast and, as they always have been, bush people were quick to adopt any new technology that shrank time and distance. In 1920 there was one motor vehicle to every 24 Australians; five years later it was one to 17. They must have been wild drivers – accident rates increased in proportion, and in 1927 there were 943 deaths and 13,569 injuries on Australian roads. The Falkiners of *Boonoke* were early motorists, and early offenders. In September 1920 Otway Falkiner wrote to Harold P Whitty, solicitor of Berrigan: 'I have been summonsed to appear at the Finley court on Wednesday next, on a charge of having driven a motorcar furiously through Murray Street, Finley, on 21 May last. Kindly attend the court on my behalf and plead guilty, at the same time protesting that the street was clear at the time and although the pace was fast, under the circumstances it was not dangerous; the car being in control of a thoroughly competent driver who has had much driving both in the country and in the busy thoroughfares of Melbourne. This is the first time I have given offence.'[1]

Otway's brother Ralph was defendant in one of the first motorcar accident cases in Victoria. The plaintiff, a newsboy, claimed that Falkiner had run into him, damaging his hands so badly that he could not use them, and causing deafness in one ear. Falkiner's defence was that the boy ran into the car, his hands were not damaged, and that he had always been deaf in one ear. The boy lost the case, but scored off the leading silk Purvis KC, who attempted to entrap him after a luncheon adjournment:

Q: *'Well, my little man, did you have a good lunch?'*
A: *'Yeah.'*
Q: *'And what did you have for lunch, my little man?'*
A: *'Steak and onions.'*
Q: *'And what did you eat it with, my little man?'*
A: *'Gravy.'*

Otway and Ralph's father, Franc Sadlier Falkiner, resisted cars at first, but according to his grandson Fred Knight later consented to ride in them and quite enjoyed the experience, but always insisted on having a rope to hang on to. For all his years of experience as a driver Otway himself never quite mastered the machine. When I knew him in the 1950s he would talk to a vehicle as if to a horse, alternately encouraging and berating it. His habit was to jam it into the first gear that came to hand, and stay in that gear all day, cursing if the car shuddered or overheated. He never trusted speedometers, and judged speed by putting a hand out the window, often shouting, 'Faster you old bitch'.

Social commentators saw moral value in widespread use of cars. An editorial in *The Pastoral Review* during the 1920s pontificated: 'The sordid isolation and the appalling loneliness which make many small families so miserable in the remote country places are gradually disappearing by reason of the ease with which long distances may be covered by the motorcar. The effect is shown in their improved mentality and in the wider interest they take in life generally. An instance was recently cited of a family living for many years under drab conditions in the outback blocks of Queensland. The children were scared of strangers, were lacking in interest in their surroundings and seemed deficient in ordinary intelligence. In due course a low-priced car was purchased. It opened their eyes to other people and other things, and it revealed an intelligence in them that they had never before given evidence of, and they now have a better chance in the world.'

Wider use of trucks improved operations on the stations. In 1924 AC Macdonald, owner of *Olive Downs* and *Naryilko* in the far-western corner of New South Wales, carted 45 rams from Broken Hill to *Olive Downs* in an International truck. The trip of 395 kilometres took 24 hours, compared to a droving journey of eight weeks. 'At least a dozen expensive rams would have been lost on that trip, and the remainder would have arrived in a knocked-up state, and have been of no use for a season.' A load of wool was moved from *Olive Downs* to Broken Hill in one day, against five to six weeks by camel team. Macdonald's stations now needed to hold only a month or six weeks' stores on hand compared to the safety net of six months when the camel teamsters had to wait until they accumulated six-ton loads at Broken Hill.

In 1927 there was great interest when WA Russell, an innovative pastoralist and member of Queensland State Parliament, purchased 10,000 full-woolled wethers at *Listowel Downs*, Blackall, and shifted them by road to *Nardoo* in the Cunnamulla district. This was a trip of 340 kilometres to save a rail journey of 2250 kilometres. It was undertaken in trucks of Russell's own design at a cost of a shilling per 80 kilometres for a truck carrying 250 sheep. Transporting sheep by truck had shifted from curiosity to reality. A new era had begun.

Through the 1920s there was fierce competition between British and American car-makers for the Australian market. Consumers were torn between their traditional loyalty to British products and the superior power of American cars – UK models were generally of lower horsepower because of the British system of taxing cars according to horsepower in their domestic market. By 1928 America seemed to be winning the battle, with Studebaker and Erskine cars showing strong growth in sales.

The versatile Model T Ford had been ubiquitous on stations since before the war, despite the lengthy procedure required for starting. It was first necessary to adjust sparking and throttle levers mounted on the steering wheel, paying particular attention to the spark – too far advanced and the engine would backfire and the crank handle could break your arm. Then round to the front, give a quarter turn on the choke handle alongside the radiator, push the crank handle in to engage the ratchet, taking care not to wrap your thumb around the handle lest you have it torn off, and start cranking. One turn for the choke, a second to build up compression, and the third to start her – if you were lucky. Otherwise you might be cranking and adjusting levers for half an hour. Then jump aboard, use the three pedals (clutch, brake and reverse), pull the throttle lever back, and you were off at a breathtaking 58 kilometres per hour. Steering was brutal and uncertain, but the vehicle was strong and reliable. Henry Ford famously offered any colour in the US provided it was black, but lucky Australian motorists could also select red or green.

In the 1920s motorists set out on long trips equipped with supplies of water and fuel, two or three spare tyres, tools, puncture kit, wire and pliers, a tucker box and enough mechanical skills to clear a fuel blockage or remedy an electrical problem. They could expect one or two punctures and a mechanical or suspension failure on every journey. Driver and passengers wore long cotton coats against choking dust as they rolled and lurched over partly formed roads still scored by the wider tracks of horse-drawn carts.

This new mobility for station travellers was not matched by improvement in country hotel accommodation. In 1925 a traveller accused the country pubs of being a blight on civilisation: 'It is scandalous that when reaching the average country hotel on a Sunday or after 6.30 pm the traveller can get nothing to eat, while even hot water is difficult to obtain. As a rule, if there are bells, you can ring until you are blue in the face before anyone answers. The lavatories are invariably disgusting, and flies as a rule swarm in the meal rooms and little attempt is made to keep them out.'[2] A higher standard was found at the Imperial Hotel in Longreach, which was described by Jameson in 1920: 'My home for the night was the Imperial Hotel, the largest. George, the porter, met all the trains in his horse cab and drove guests and baggage to the hotel, part of the service. The Imperial was owned by Mrs Springer and was run on strict lines; she insisted that all males in the dining room wore a coat, even if the temperature was over the 100 degrees Fahrenheit mark. If you had no coat you could take your pick of a number of washable coats hanging in the office, and some of the old diners looked characters in those coats. The meals were very good, cooked by a Chinese cook with quite a reputation.'[3] The Imperial, a fine old weatherboard structure, burned down after the Second World War in mysterious circumstances. The local member of state parliament, a bachelor, lived in the hotel; when woken in the dead of night by the house boy with news that the hotel was on fire, he was reputed to have rolled over and said, 'Don't be silly, that's not until next week'.

In 1921 one era was closing as another opened. Cobb & Co. announced its last coach service in Queensland was to close, while at the same time air travel was becoming possible. That year Queensland people were amazed when an aeroplane enabled a buyer to travel out 160 kilometres from Longreach, inspect 28,000 wethers, return and complete the deal all in one day. Western Australian Airways ordered construction of four De Havillands at £60,000 for their Perth–Adelaide service in 1928, claiming a cruising speed of 110 mph, and reassuring customers that each machine had three engines, and if two petered out it would still be able to fly 50 kilometres. Flying certainly had its risks. In 1921 it was reported that British civil aircraft had flown 1,556,000 miles (2,500,000 kilometres) since 1919, carrying 106,712 passengers. They had suffered 48 crashes, of which 20 involved deaths or injury. Nervous passengers would have been little reassured by the boast that the ratio of passengers killed was only 10 per 1000 carried.

In 1920 the Larkin-Sopwith Aviation Company was advertising the Sopwith 'Dove' biplane with 80 hp Le Rhone engine for £1350. Maximum speed was 153 kilometres per hour, petrol consumption six gallons per hour, capacity 16 gallons, and they claimed it was 'a most satisfactory machine for all classes of pilots, and particularly useful for the businessman'. For the sake of flying woolgrowers, one hopes it revealed more placid habits than its famous wartime cousin the 'Camel', which was described as 'a moody, savage beast. Some days it would perform all the stunts, on others none. Sometimes it would touch 120 mph, and on others just 70 mph, and the pilot would return black as a pit pony, drenched in evil-smelling castor oil, the Camel's lubricant. At any time petrol vapour might be drawn through the exhausts and become flaming streamers which threatened to seize on the fabric covering. Difficult to land, deadly to crash, the first flight in a camel was a terrific ordeal.'[4]

While petrol engines were increasing speed on and over the ground, steam was still being used on the stations to drive shearing machinery and pumps. Despite their insatiable hunger for wood, there was something about steam engines that people loved – perhaps it was because each driver could coax something different in performance from an engine, and each had different characteristics. Cameron at *Barratta* revealed his affection for a favourite engine in a letter to head office in 1922: 'The old steam engine worked satisfactorily through the shearing, and expert Lumby who drove her during shearing was satisfied she was a very good engine, and capable of doing the work required of her. I think if I have her well overhauled it would be a fair risk to use her again this season. It would pay us to then keep her here for pumping water and such work, which she enjoys.'[5] A steam engine under full power was an awesome sight; glistening with heat and moisture, rocking on its steel wheels, hissing, spitting, and blowing off steam and smoke. The huge cast-iron flywheel was watched carefully, because if the governors went out of action and the engine bolted, it could disintegrate into a lethal shower of shrapnel-like fragments.

As fearsome as steam engines may have been, their oil-driven cousins could be just as brutal. A Phoenix oil engine was used at *Byra*, near Barellan, New South Wales: 'Her habitual roar as she worked was terrifying. Only one man seemed able to pacify her. In the full strength of her youth it took, usually, about a dozen men to start her. Once a shearing contractor's man undertook to work her, and all went well until shearing was nearly finished, then with a hoarse splutter, which seemed to wrench her innards, she playfully tossed her huge flywheel into about a dozen pieces, hurling the fragments among the 25 shearers, nearly killing the ringer, while the rest took refuge under the shed. A new flywheel was procured, but the shearers' nerves were quite gone, and when someone accidentally dropped an empty petrol tin one day, all hands dived to earth.'[6]

As the 1920s wore on, time and distance were certainly shrinking, and there was great optimism in the bush about the way things were going. Jameson summed it up when recalling 1922: 'Although not in the very early days, the period when I commenced jackarooing at *Westlands* was a very interesting one. I had seen the last of the bullock teams and virtually the last of the horse-drawn wool teams; motor transport was taking over on both the roads and stations. Also at the same time Qantas was being born: Hudson Fysh was flying around in his small plane trying to form his airline.'[7]

The 1930s saw heavy draught horses being replaced for farming operations. The number of tractors in South Australia, for instance, nearly doubled from 2503 in 1927 to more than 4000 by 1930. The Clydesdale horse, with its gentle nature, deep red colour and white face, was the noblest of all draught animals. During the 1900s the Falkiner family ran one of Australia's leading Clydesdale studs at *Widgiewa* and, to stop the mares from putting on too much weight, would yoke them up to the wagon that met the daily train from Sydney. Such was the general appreciation of good horses in those times that the train would stop to allow passengers the pleasure of seeing the team plodding its placid way, with foals trotting alongside. By 1938 there was concern for the future of the Clydesdale when four studs were shut down in South Australia. By 1939 there was fear that the decline in draught horse numbers might inhibit the war effort and, if not arrested, agriculture could break down. There was a call for subsidies for wheat that was grown under horse power.

The reliability of cars was improving, and they were becoming more available. In 1930 a Willys Overland Whippet Six, with oversize balloon tyres and 'finger-tip control' through a single button for starter, lights and horn, was available for £375 the sedan and £285 the tourer. The Austin Motor Company received excellent publicity when their tourer, with gears sealed in top, made a record run of 13 hours 20 minutes from Melbourne to Sydney, averaging 68 kilometres per hour. At the close of the decade canvas hoods and square bodies gave way to a more streamlined look, such as the Humber Super Snipe that was available in 1938 at £498. But motorists still needed to be prepared for all emergencies when driving on country roads. When journalist Lionel Krone set out in his giant Crossley car to tour the Riverina in 1933 he fitted patent double-springs to the front suspension, loaded an axe, shovel, two extra tyres and tubes, chains, thermos, gum boots, and a 'Comet' winch for pulling the car out of bogs.

Travellers were also still complaining bitterly about the standard of service and accommodation in country hotels. In 1936 one called attention to: 'The filth, discomfort, and inattention – the only latrine is to be found at the far end of a noisome yard, redolent with rank odour of stables, food waste, surface drains and slops, yet not pungent enough to subdue the smell of the place to which every guest must on occasion find his way.' Another told of how one hotel had installed a smart new bath: 'Quite an up-to-date article of cast iron with white enamel finish. One guest, on failing to find the bath in its usual place one evening, went in search and found it in the back yard, the centre of attraction for all the family and a few helpers who were using it to scald a pig that they had just slaughtered.'[8]

In 1937 Glasson Motors in Broken Hill took delivery of the largest stock-carrying vehicle in the Commonwealth – a 48-horsepower Leyland six-wheel truck pulling a trailer 16 metres long, three metres wide, on eight wheels shod with 10-inch tyres, and designed to carry 350 sheep. The cab was fitted with a two-way wireless set. Descendants of this vehicle were to completely revolutionise the trade in sheep following the Second World War. But the new technology had its downside, and in 1939 there were reports of substantial thefts of sheep on the Darling Downs – thefts on an organised scale by means of high-powered motor lorries in the dead of night.

In 1930 air travel was becoming quite commonplace. Australian National Airways was offering daily flights from Brisbane to Sydney for nearly £29 return. Flying time was six hours. Qantas had flights five days a week to Longreach, Camooweal, Cloncurry and Normanton. In 1936 Reliable Air Travel Limited was floated with capital of £20,000 to establish a commercial service between Brisbane and Cunnamulla and towns en route. In 1939 Queensland pastoralist CW Russell purchased a Beechcraft plane from America capable of 300 kilometres per hour for travel between his various properties.

For the affluent, overseas travel was becoming less time-consuming. By 1936 it was possible to reach London from Sydney in 28 days taking the liner *Mariposa* to America, travelling cross-country by rail, then sailing on the Cunarder *Queen Mary* across the Atlantic. In the mid-1930s the grazier Norman (later Sir Norman) Kater undertook, as part of a leisurely England–Australia trip, the hop by air from London to Capetown for £130 including all meals, tips, hotel accommodation and transfers. He reported that the Handley-Page *Helena* aircraft 'had four Bristol Jupiter engines of a combined horsepower of 2200, no vibration and no noise, a bar and steward, a spare pilot, two-way wireless, and roomy luxurious conditions for 38 passengers.'[9]

During my career, radio was to have a more profound effect upon station life than any other improvement in communication. It brought education, market reports, cricket scores, racing results, and local and world news to lonely homesteads. Later, reliable two-way technology gave access to medical care and the school-of-the-air, and was to revolutionise the way big properties were operated. A Longreach district jackaroo recalled: 'Wireless was just coming in. In 1929 the jackaroos clubbed together to buy a radio. It was a huge thing, and not very successful, but later people rigged up crystal sets. We had a wind-up gramophone out in the mustering camps, and whenever a hawker called we would buy more records.'[10]

Wartime petrol rationing was introduced in October 1940 and, like food and clothing, was controlled through the issue of coupons. A Liquid Fuel Control Board was given absolute control over the sale and use of all fuel, and vehicles were divided into 25 categories according to priority of use. For example, a vehicle of 12-horsepower used for the carriage of passengers for private and business use, that had travelled between 6001 and 8500 miles (9660 and 13,680 kilometres) the previous year, was designated class three, and was entitled to sufficient fuel to travel 390 miles (627 kilometres) per month. By 1942 it was compulsory to fit gas producer units to all vehicles. These produced carbon monoxide gas by passing a flow of air over or through a bed of red-hot charcoal. Engines were started and run for three minutes on petrol, then switched to gas. There were many problems, because the gas producers were temperamental; they would run efficiently one day and refuse to function the next. Apart from the loss of 40 per cent of normal power, drivers had to be conjurers to adjust to the right proportion of air to gas. Suspension and steering were affected by the 300 pounds added weight of the unit, an air leak in any part of the system would give real trouble, and clay or dirt in the charcoal could cause formation of great rock-like 'clinkers' that blocked air supply. Sometimes charcoal failed to trickle down into the fire zone due to formation of an arch of material in the hopper, causing loss of power. The solution was to give the whole unit a good shake, so drivers could be seen suddenly spearing off the formed road without warning to drive a few minutes over rough ground. The drum-like metal body of the

The Imperial Hotel in Longreach, Qld, offered a high standard of comfort and service in the 1920s *(Longreach Powerhouse Museum)*.

producer often glowed red hot, and many was the piece of luggage burnt to a cinder in the back of utilities.

Wartime shortages and rationing disrupted most business operations. In 1942 wool brokers pulled their travellers and stock salesmen off the road, and all business was done by phone. That year Western Australian graziers ran out of rams because local supplies had dried up, and rams could not be sent from South Australia as the transcontinental rail line was being fully utilised for military purposes. Northern pastoral areas could not even get local supplies because shipping out of Fremantle was not available.

By the end of the war most people were still driving much-repaired 1936 model cars and utilities; particularly utilities because banks would not finance purchase of sedans but supported purchase of utilities which could be used for both work and social purposes. Keeping these vehicles on the road tested the ingenuity of bush people, because replacement parts were generally unobtainable. In February 1944 a northern grazier complained: 'An accident to its radiator put a utility out of action and it was four weeks before I received a new one. It was eight months before I received two tyres and tubes. Then a battery on a lorry did the usual thing. It conked out on December 20, 1943 and up to date I have not got a new one.'[11] As a boy I saw my father fashion a new distributor arm from a mulga branch using a pocket knife, and vehicles were seen around with all sorts of strange arrangements of wood and wire replacing bumper bars, doors and mudguards.

Progressive relaxation of rationing and other restrictions began to free things up when the war finished. In 1947 Blue Bird Air Taxis were offering services out of Bankstown aerodrome with their newly imported Austers, which carried two passengers at the rate of 9 pence per mile. They claimed these aircraft to be the 'most easily manoeuvrable aircraft yet built, having such low flying speed that stalling is virtually impossible and side slips are eliminated.' In 1949 Trans Australian Airlines announced they would be using De Haviland Drovers, all-metal machines with three motors carrying five passengers, to link all western Queensland towns from Longreach. The schedules included Barcaldine, Aramac, Muttaburra, Stonehenge, Jundah, Windorah, Eromanga, Quilpie, Thargomindah, Charleville and Adavale. This was a better service than that available to the outback 50 years later. Also in 1949 British Commonwealth Pacific Airlines were advertising full-length curtained bunks on their 483 kilometres per hour pressurised DC6 service, from Sydney to San Francisco in 29 flying hours. The world was getting smaller.

The late 1940s brought the biggest change of all in terms of bush transport. Four-wheel-drive Jeeps, ex-army and left-hand drive, were flooding into the bush. These were the first truly off-road vehicles available to the stations – for the first time town became accessible in almost any weather, and people began to realise sheep could be mustered and handled adequately using vehicles. This would set the scene for the demise of the station horse and huge staff reductions, stripping some stations almost bare of people and decimating many little hamlets.

In 1946 General Motors gave the first public display of their new, locally made Holden car, with six-cylinder motor, 'Aerobuilt' body and ample head and leg room for five or six. They claimed: 'You'll find comfort and safety, high speed and easy handling. You'll find life, action and thrills.' With brutally direct steering and drum bakes that could grab and flick you over in a wink, the thrills were maybe too frequent, but it was the opening of a new era. At £675 plus sales tax the car was good value rather than cheap, but the convenient hire-purchase terms available through their own finance company made cars accessible to many more people. The grip of isolation on station employees – because of their dependence on station transport – was being broken.

Yet even the most beneficial change will have its critics. One back-country man, accustomed to the old practice of stopping for a nip at the wayside shanties that in early times existed every 15 to 25 kilometres, wrote in sadness: 'Journeys that took days by coach or buggy are now made in a few hours by car, and under infinitely more comfortable conditions. This rapid transport has, sadly, sounded the death knell of the wayside pub. Such dry stretches as Wanaaring to Wilcannia and thence to Cobar, a distance of more than 300 miles, are indeed anathema to we thirsty souls.'[12]

At the same time other things were happening that would change forever the face of inland Australia. Large quantities of surplus US military heavy machinery salvaged from the Pacific theatre of war were becoming available. Huge Caterpillar tractors and bulldozers, many still encrusted with barnacles following immersion in tropical seas, got busy clearing vast areas of scrub and sinking dams in Queensland so that more Merino sheep could be run. Fifty years later the environmental jury may still be out on this activity, but it did much to generate export income for Australia's post-war recovery, lifted standards of living, and enabled the nation to support a larger population. A more modest piece of machinery was also invading stations – a host of little grey tractors. The Ferguson TE series was made in England by a division of Standard Cars, who patented a unique hydraulic three-point linkage system which meant it could pull a plough that previously could only have been handled by more powerful machines. Like Model T Fords in the 1920s they became ubiquitous, doing every job from ploughing and carting rubbish to transporting people. The use of farm tractors and their hydraulic systems was to take away much of the heavy lifting and bullocking work that were a common part of station life in my father's era, and which ensured that every farm worker ended up with lower back problems. But the chief function of Ferguson tractors was controlling rabbits. They trailed a two-prong ripper, which could plough rabbit warrens down to a depth of 60 centimetres, trapping the occupants inside and suffocating them. It was to be the first truly effective low-cost weapon against this pest, which was taking Australia's rangelands over and represented the single greatest threat to the natural environment – a greater threat than scrub-clearing or European farming methods has ever been. It was only the intervention of such technology and a set of fortuitous circumstances during the 1950s that avoided inland Australia's vegetation being decimated and topsoil carried away by the westerly winds.

CHAPTER TEN

COMPANY MEN

No man is born into the world, whose work
Is not born with him; there is always work,
And tools to work withal, for those who will:
And blessed are the horny hands of toil!

JAMES RUSSELL LOWELL (1819–1891)

They say never go back, and they are right. During 1997 I returned to the site of my childhood home, *Buckiinguy*, nestled on the banks of the Macquarie River 100 kilometres north of Warren in New South Wales. Relegated to an outstation in the 1950s, partly demolished in the 1960s, the neglected and empty shell had finally burned down following a lightning strike in the 1970s. An untidy pile of rubble was all that remained. About the only recognisable things were a bougainvillea and the aged and decrepit wooden garage, leaning perilously as it awaited its own demise. Gone to ashes were the cool gauzed verandahs, the date palm that never bore fruit, the lovely Jacaranda trees, the unstable tank stand, and the snake-infested cellar. And all the happiness and security the house had once contained.

Everything had shrunk. The raised path down which my father made his nightly torch-lit odyssey to shut down the lighting plant, the dusty track to the shaded mysteries of the blacksmith's shop – which were vast journeys to a child – were foreshortened, insignificant, and out of all perspective as I remembered them. The access road was crumbling away, culverts falling in on themselves. Soon the remains of the old homestead would be absorbed by encroaching bush and be lost to human sight, save for some pig-hunter who may stumble across them as he seeks his game across the Macquarie Marshes.

Back in 1931 *Buckiinguy* was a bustling sheep station, alive to the morning sounds of galloping horses, bells, the clash of stirrup irons, cracking whips, barking dogs, and the shouts of stockmen. Its 34,000 hectares were carrying 34,237 sheep, 690 cattle and 85 horses. The staff consisted of the manager, JA Morrison, overseer M Murdoch, just appointed after five years as a jackaroo, and 16 men. The Australian Estates, one of the great Australian pastoral companies, then owned the property. My father was manager there from 1925 to the early 1950s, and the memory of the place still haunts me.

For the best part of a century eight major pastoral companies owned strings of sheep stations right across eastern Australia. They were: Australian Estates, Scottish Australian Company, New Zealand and Australian Land Company, Goldsbrough Mort, Australian Mercantile Land and Finance Company (A.M.L.&F.), Australian Pastoral Company, Squatting Investment, and Australian Agricultural Company. Most had their principal shareholders located in the United Kingdom, and began their lives during the 19th century as investment companies lending funds in the Australian pastoral industry on first mortgage security. They became property managers later as a matter of necessity. They were big, well-resourced and efficiently managed. They took care not to overstock their land, bred high-quality flocks and herds, and provided commercial leadership to the rest of the industry. They set the benchmarks.

A history of the Scottish Australian Company written in 1963 recalled: 'The Scottish Australian Company was founded in Aberdeen, Scotland in 1840 for the purpose of acting as an investment company. Most of the 400-odd shareholders were residents of north-east Scotland and included representatives of every class of the community, from Lairds or landholders to labourers, fishermen and shopkeepers. The company, during the depression of 1843–44, had come into the possession of much pastoral land and thousands of sheep in New South Wales, but it had always sold those as quickly as possible or, where this could not be done satisfactorily, leased or let them out to carefully selected people at a good rental.'

Bank closures and the devastating droughts of the 1890s and early 1900s forced the companies to take possession of many properties, whose owners just packed up and walked off. The Western Division country of New South Wales had been virtually demolished by the drought, losing most of its surface vegetation and much of its topsoil. Of the 308 pastoral leases in the division in 1900, more than 100 were registered in the names of pastoral finance houses and 65 in the names of banks. During the 1890s A.M.L.&F. Co. foreclosed on 19 of the 32 in the region who were indebted to them. Of the nine indebted to Australian Estates, foreclosures had overtaken seven by 1905. By 1900 Goldsbrough Mort had acquired 12 stations in New South Wales, the A.M.L.&F. 20, and Australian Estates nine.

Inevitably there were problems. The owner of *Canonbar* and *Meryula* in the Nyngan–Cobar area surrendered the management in 1895. 'The inducement was an allowance of £625 per annum, upon which he divided his time between Sydney and *Canonba* where he maintained, at the general station expense, a separate establishment away from the homestead and used his legal rights to harass the management (put in place by Goldsbrough Mort) at every turn.'[1]

Despite the human misery caused by the foreclosures – there were many who would have seen them as a merciful release – it was in the long run in the best interests of the country and its natural environment. As HM Cain pointed out: 'The significance of growing company influence in the west was that, at a time when knowledge was imperfect and its communication was slow and inaccurate, the bringing together of many stations under one guidance facilitated the development of a more effective relationship between the pastoral enterprise and its arid environment.'[2] Most of the companies were good husbands to the land. They stocked conservatively, restored the ecology, refurbished and made structural improvements, and built up high-quality flocks and herds.

The men who managed these properties for the companies bore heavy responsibility, thought of little other than their work, and were usually poorly paid. In 1932 my father was managing the *Buckiinguy* run for a salary of a little more than £300. Little changed over the next 35 years – in 1967 the manager of *Amby Downs* at Mitchell in Queensland for the Australian Pastoral Company was receiving, in terms of value, a similar amount. The old pastoral companies were never over-generous in matters of salary. One jackaroo wrote in the 1920s: 'I had been due for a rise in wages after two years experience, so I informed the bookkeeper Gordon Teasdale, and received the additional amount. When the returns got down to head office it was pointed out that during the two years I had taken a holiday of three weeks and I would not be entitled to the additional amount for another three weeks.'[3] By 2000 things had improved, and the responsibility and workload taken on by station managers were more widely recognised. The manager of a large station was being paid in the range of $40,000 to $50,000 which, considering the tax-effective nature of housing and perquisites received, was equal to $90,000 to $100,000 for a city-based manager of a medium-sized business.

Though poor in salary terms, pastoral company managers enjoyed many additional benefits. There was usually a substantial furnished homestead in an established garden, and the English companies expected homestead life to be maintained with a certain style. Up until the Second World War they provided domestic staff: a cook, perhaps two housemaids, a houseboy, gardener and groom. Dinner at night was formal, with the overseer, bookkeeper and sometimes jackaroos included, and during the 1920s and 1930s some companies stipulated that evening dress should be worn.

On a few stations the manager's wife preferred to have the main meal in the middle of the day, which could be uncomfortable during summer. The Broad family at Australian Estate's *Raby* station followed this practice, and I shall never forget a visit there as a child in the 1940s. It was a searing summer's day, and the heat radiated into the heavily curtained dining room. Jim Broad at the head of the table was carving a great steaming leg of mutton, and Mrs Broad served baked potatoes, pumpkin and creamed cauliflower. A rich plum pudding and hot custard followed. They were still using gas lighting at the time; the gas hissed and spat, as everyone sat around the table wearing heavy tweed sports coats over their working clothes and streaming with perspiration. It was like a scene from *The Inferno*.

During the hard years of the 1930s and 1960s a station manager and his family often enjoyed a more comfortable lifestyle than most property owners in their district, and they held considerable social status. The station was usually one of the biggest businesses in the district, and the manager wielded substantial power through his buying and selling activities, so local suppliers courted his favour. Most of the companies encouraged their managers to take a role in community affairs. 'This was partly motivated by a desire to enhance the public standing of the station; partly to encourage the development of community amenities and so improve conditions for those who worked on the stations; and partly to see that the shire rates were kept as low as possible.'[4]

Though mean-spirited in many ways, the companies enjoyed great loyalty from their staff, and people were proud to be identified as 'Estates', 'Land Company' or 'SACO' men. This came about because the companies took a close interest in the lives of their people and their families, and could be magnanimous in times of trouble. On one occasion Australian Estates general manager GS Colman, on his annual tour of the stations, noticed my father was unwell. On return to Melbourne some weeks later he sent word that the whole family should take a six-week holiday at the expense of the company. My father suffered a good deal of ill health during the early 1950s and underwent a series of major operations. This was at a time when there was no properly structured health insurance available, so my parents were battling to pay the bills. I was mid-way through secondary education as a boarder at Scots College in Sydney, and Australian Estates offered to either pick up all the medical bills or pay for my schooling. My father chose the latter. Later, the Federal Repatriation Department decided his health problems resulted directly from First World War injuries, and paid all medical costs retrospectively. When he advised the company, they elected to continue to pay for my schooling, and then offered to send me on to university.

When Murray Clapham, then general manager of Australian Estates, died at the age of 40 in 1925, he had no superannuation. The company handled all his widow's financial affairs for the next 50 years, and settled £1000 a year on her for life. Later, most of the firms would set up employee superannuation funds, and during the good years ploughed a share of their profits back into the schemes. Following my father's death in 1957, my mother – who had lived most of her married life in a state of genteel poverty – enjoyed a long widowhood of modest affluence, thanks largely to Australian Estates generous superannuation scheme.

Despite the financial security offered by these schemes, retirement for the old-time station managers could be a difficult period. In modern times they tend to retire to the large regional centres, such as Dubbo, where they can keep in touch with acquaintances. But in my father's time people went straight to a capital city where they were completely cut off from the world they knew. Used to a vigorous outdoor life, many hated the city. I can remember my father sitting for hours in our little house in Chatswood, just watching the traffic in Fullers Road, until in desperation he took on relieving manager and bookkeeper jobs to get out of the city.

In bountiful years most of the companies paid bonuses to their station staff. During the 1950s it was the practice of Australian Estates to present these in an envelope at Christmas dinner. Rowand Jameson recalled that at the Australian Pastoral Company in the late 1940s: 'Everyone who had been with them for six months received a generous bonus, and the managers a much larger one. Mr Young (general manager) said, "There's no better time, Jameson, to give the managers a pep talk in regard to their shortcomings in management than when you are handing them a good fat bonus."'

The companies developed a tough, demanding, but family-like internal culture. They treated managers with dignity, and many spent a lifetime in their employment; often sons followed fathers. James Mitchell, who managed *Wellshot* in the 1930s, had lived under the New Zealand and Australian Land Company umbrella all his life. In 1986 he recalled: 'I was born in 1896 on *Bangate*, a company property near Walgett, which my father managed. I suppose my working life with the company started when I began running in the horses on *Bangate* when I was eight years old. I went to the First War and when I came back the company gave me the sub-overseer's job on a place that adjoined *Bangate*. Then I was appointed manager at *Maxwelton*. I had that job for a year before being appointed manager at *Wellshot* in 1930.'[5] My father's career ran on similar lines. He joined Australian Estates as a jackaroo straight from school in 1913, worked at *Brookong*, went to the First World War from there, then on his return worked at *Thackaringa*, *Cambridge Downs* and *Terrick Terrick* before landing the job at *Buckiinguy* in 1925, and was still working as a manager for the company when he died in 1957. I too began my career with Australian Estates, but ambition for early promotion made me restless and I moved on. It's as well I did, because the company was to be broken up within 20 years. Australian Estates, like many of the companies, was very conservative, and in the 1930s and 1940s seldom sacked a manager unless malfeasance was involved. If his performance was not up to scratch they kept transferring him, with no loss of salary or perquisites, to jobs of ever-diminishing importance. By the time he ended up at one of the stations that ran only wethers he was expected to have got the message and resigned. The company also expected their managers and executives to observe a moral code. During the 1930s one promising young man based in Melbourne head office was quietly 'disappeared' from the company when, on a visit to one of the stations, he was caught in a hot embrace with a housemaid behind the kitchen.

The station managers were intensely practical people, who wrestled with practical problems. Much of their tactical planning was preventative. They were engaged in working with nature, but at the same time waged a bitter war against nature in her crueller moods, and were constantly trying to stay one move ahead. Hence they would anticipate a flood and move sheep to higher ground, jet sheep before a fly-wave struck, move stock around pastures in a gigantic chess game aimed at keeping them in good condition, move early to give supplementary feed, and lighten numbers before drought took its toll. To the uninitiated this work was unspectacular and often went unnoticed; it only became spectacular when a round was lost.

For most of the 80-odd years covered by this book, the financial performance of a station depended upon the manager extracting maximum productivity from his livestock – heavy fleece weights, high lambing percentages, good bodyweights in sale stock, all accompanied by low mortalities. As John Ryrie Graham said of sheep in 1870: 'Feed them well, and they will pay for it; half-feed them, and they will not half-pay you; neglect them, and they will ultimately ruin you.'[6] Stock required constant attention. In a column of advice in *The Pastoral Review* during

Wanganella Estate Homestead, Deniliquin, NSW, before it was burned down in 1935 *(F.S.Falkiner archives)*.

the 1920s 'Old Hand' wrote: 'In our own time we have certain rights, but in daylight, early or late, Sunday or a week day, it is a crime to neglect stock. As a matter of fact a station manager has no time that he can really call his own.' He recalled some advice he had once given to a friend: 'I told my friend that the whole of his time is paid for, and that he should never be at home during daylight so long as there was need of his presence on the run.' Indeed, much of a manager's time was spent patrolling his run, checking livestock, water and pastures, and planning his next moves. One wrote of his work in the 1940s: 'I would be away from the homestead for the best part of a week at times. I spent a good deal of my time in the utility, plus gas producer, swag, tucker box, plenty of water, portable telephone, and pull-out gear in case of being bogged.'[7]

By the start of the 21st century, the standard of sheep husbandry in Australia had deteriorated. The skills of the stockman were being lost because there was no-one around to teach them and most young people found sheep boring anyway. Reducing costs had become the most important component of management, and the shift from the former man/sheep ratio of, say, one to 1500 up to one to 10,000 or more ensured sheep could not receive the same level of care and attention. There was just insufficient time. The type of sheep bred to cope with this environment will probably revert to the old easy-care, light-producing types that were around in the early days of settlement.

The work environment on the old stations built fierce commitment and dedication. It was said of one manager at *Beaconsfield*, near Longreach: 'Anderson's dedication to his job and his disinterest with anything beyond the bush were such that he only visited Longreach once in 13 years. When he did so, he complained bitterly that the banks and business houses were not open by 6 am.'[8] Such distaste for town and society was not confined to the senior ranks. During the 1960s there was a gardener at the Falkiner property *Zara*, near Deniliquin, named Charley Spalding. Charley was a splendid fellow who just got on with his work in an unfussed way, uttered few words, did not own a car and had little interest in going to town. However, there was a strict rule on the station that each year the overseer should take Charley for a day in town on the closest shopping day to Christmas. One year it rained heavily, the access road became impassable, and the contrite overseer had the unpleasant task of going to Charley to advise him the trip was off. 'Don't worry Bob' was Charley's philosophic reply. 'Next year will do.'

The old station managers adopted many different operating styles. Most were autocratic, a few were consultative, some worked hard physically alongside the men, most adopted the 'patrol and monitor' style, and a very few were stay-at-home 'verandah managers'. In early days the style was strictly military. James Miller managed *Portland Downs* near Longreach from 1888 to 1897, and was described as a 'grand old man. The ordinary men were handled "Army Fashion". A bugle would sound the "Fall-in", the overseer would report "All present Sir", then the boss would set each man his day's work.'[9] Men like my father, who had been officers or NCOs during the First World War, could be quite physical when dealing with staff. He was the most gentle of men with his children, but I can remember a former jackaroo telling me he was 'a fiery little bugger, who would shape up to you pretty quickly if you did the wrong thing'.

Some were of a religious bent. There was a famous manager in the Riverina whose custom it was to conduct a church service in the homestead for shearers and shedhands during shearing. He would conclude the service with a prayer: 'O Lord Thou knowest that in these hard times wool is wool and that a few ounces per sheep makes a great difference to the total clip. Put into the hearts of these Thy shearers to keep down to a fair thing, and O Lord I would beseech Thee to keep Thine eye on James Bourke and William Brown, and O Lord I would further beseech Thee to put a stop to the awful tomahawking that went on during last week. Amen.'[10]

In addition to their outside duties monitoring livestock, pastures and daily work projects, managers were responsible for a good deal of administrative work supplying various returns to head office. I recall my father having dinner, listening to the ABC radio news at 7 pm, then disappearing to the office until 10 pm or later. During the 1930s the manager at *Barratta* for the A.M.L.&F. Co. was required to supply a comprehensive monthly report to head office. It included a list of livestock on hand by age and category, and a reconciliation of stock movements for the month – lambs marked, purchases, sales, transfers and deaths. Also a summary of the season, the outlook, condition of stock, a copy of all ledger sheets and balances, a list of employees and wages, particulars of outstanding contracts, and a summary of operations for the month. He was required to provide every six months a complete inventory of stores and materials on hand and their value, an inventory of plant and furniture, and the value of each item. Also shearing and crutching returns year to date, a series of indicators which included performance against budget and total cost of wages and rations per head of sheep, and a list of plant renewals and repairs and maintenance carried out for the period. The annual report on operations of *Buckiinguy* for 1931 was comprehensive, and its scope and level of detail is not exceeded even in the current age of computers. The 29 typewritten pages included a list of land titles and areas, a profit and loss statement, balance sheet, and analysis of expenses. Also summaries of depreciation, livestock on hand, stock movements for the year, and lambing and shearing returns, all with comparisons to previous years, and rainfall for the year with 10-year comparisons and a narrative describing the current season. In addition, there was a list of vehicles and plant, improvements carried out during the year, a report from the manager with his recommendations and summary of future prospects, and the pastoral inspector's report on his two visits.

Some companies, such as Australian Estates, tended to leave tactical decision-making entirely to their managers once annual operating budgets were approved, and interfered little so long as budget targets were met. Budgets were prepared in great detail, down to the last few tins of jam to be used, and woe betide the manager who overspent his budget, for then telegrams would fly like hail. Others, like the Australian Pastoral Company, were more interventionist. The AP Company owned *Noondoo* at Dirranbandi, still 260,000 hectares in the 1940s, and another six stations in Queensland and one in New South Wales. Rowand Jameson, who was their pastoral inspector in the 1940s, described their style of micro-management: 'Mr Young (general manager Euston Young) ran the company to good effect and in the minutest detail, but was not easy to work under; with everything done to a set pattern it left not much room for original thought or initiative. I felt sorry for the managers in this respect … they were directed in all their work, and in some ways were like overseers.'[11] The directors of the AP Company all resided in England; they had appointed Gibbs Bright and Company as their managing agents in Australia, and the general manager had no direct contact with the board. He dealt with Gibbs Bright in Melbourne; they received copies of all correspondence and managers' reports, which were incorporated into massive minute books that were periodically sent off to London. All stores for the entire group, except for perishables that were purchased locally, were ordered through Gibbs Bright on the basis of six-monthly requisitions, and supplied in bulk.

Managers and head offices of all the companies maintained a constant warfare on small costs, and coded telegrams and 'please explains' flew back and forth, querying the cost of items and budget over-runs. In 1922 *Barratta* manager CJC Cameron wrote to the A.M.L.&F. head office: 'I think that the local storekeepers' charges are excessive and the goods supplied mostly of poor quality ... We get 25 per cent off for monthly payments. I advised Emry Stores some time ago if they could not supply with better quality and at reasonable prices I would be obliged to give someone else a trial. It was of no avail. I am therefore giving Douglas Bros our orders but am afraid they will be no better.'[12] As late as the 1960s many stations were only settling local supplier's accounts every six months, some only once a year. In 1968 the then AP Company general manager WS Capper was castigating his managers for holding too much inventory: 'It is bad business. It means money lying idle. For example it is not necessary to be carrying the following: 100 cakes of toilet soap, 21 tins of Sal Vital, 20 Large Pie Apples, 55 packets Lux, 24 bottles Handy Andy, 328 packets of cigarettes. Would you please watch this aspect of your store?'

Most of the companies employed pastoral inspectors who visited the stations half-yearly. The managers took these inspections very seriously, because their performance was being measured. Duncan McPhail was chief pastoral officer for Goldsbrough Mort based in Melbourne, and during the 1940s he inspected *Corinda* at Aramac in Queensland. At the time *Corinda* was running 80,000 sheep, 5000 cattle, and had paddocks up to 15,000 hectares in size. Station manager Rowand Jameson described the inspection: 'When Mr Mac had breakfasted he sat on the verandah and rolled his ration of cigarettes for the day, then went over to the office and looked at the day's inspection set up on the map spread out on the desk. In the meantime the utility was filled up with benzine in case the gas producer played up, and the back was filled up with plenty of charcoal in bags, with lunches, billy and tea in the tucker box. When the gas producer warmed up we were off for the day. Each water improvement was inspected, the stock in the paddock, the pasturage, sheep and cattle yards and the horses. We had many discussions and some arguments as we drove along, and he insisted on taking his turn opening gates. At that time he would be in his 60s, thickset, spoke slowly and little, but he did quite an amount of thinking and had an eye on him like an eagle which missed nothing. He wore a dustcoat and I can still see him walking round with hands in both pockets.' As usual before an inspection Jameson had made the place shipshape, repaired many gates and painted them white. 'Mr Mac approved of the white gates but was not over-impressed, as they grew no wool or beef and would not add to the revenue. As we drove along Mr Mac was firing all sorts of questions at me, such as how much it cost to land a tin of jam at *Corinda*. Each night before dinner we had a couple of drinks; Goldsbrough Mort were generous to their managers and sent in a case of McCallum's whiskey about every two months. Discussions continued over the drinks and often again after dinner, and the big questions came up. The inspection was coming to an end. We had covered more than 1000 miles, all done with a gas producer, six days in the utility and the seventh in the office.'[13]

On the Australian Estates stations the annual visit of Melbourne-based general manager GS (Later Sir Stanley) Colman was treated as if royalty was calling. Colman had managed a meat exporting company in Queensland before moving to run Australian Estates in 1931, where he remained until his death in 1966. 'GS', as he was known (though he was always addressed as 'Sir'), was a sparely built man with thin features, a prominent nose and delicate hands. His carriage was erect, his manner courtly and his general appearance austere. For a fortnight before his visit, the station was cleaned and polished, rubbish carted away, watering points tidied, broken rails repaired, vehicles washed, paintwork patched. An aid-de-camp would arrive a few days before to check that everything was arranged to Colman's liking. Managers and overseers wore ties during his visit, even if it was at the height of summer.

One day in 1953, when I was a raw young jackaroo at *Terrick Terrick*, manager Ray Murdoch was making a final inspection before 'GS' arrived, and the only untidy thing he spotted was me. So he sent two of us out to a back paddock de-silting a tank and out of harm's way. It was summer and we had no washing facilities, only a 44-gallon drum of drinking water, so we soon degenerated to a filthy and unkempt condition. However, Colman had recently seen my father, then at the company property *Tyrie* near Trangie, and insisted on catching up with me – he was punctilious about such matters, and generous to a fault with his time when dealing with staff. We had no warning; the first we knew was when Colman, neatly dressed in sports clothes, hove over the banks of the tank surrounded by Murdoch and various hangers-on. My companion saw them first. 'Jesus! Here come the toffs,' he cried, and ran off to hide in a nearby patch of scrub. Colman chatted for a while and relayed messages from my father, but his nose did twitch a little. It was a nice gesture, and typical of some of the old companies.

The AP Company was less understanding when in 1967 RI Balmain, manager at *The Well* near Dirranbandi, wrote to general manager Capper seeking to start shearing early so he could get married before the onset of Lent. Capper remained unmoved, and replied: 'I cannot see shearing being in the way; the sheep will go off the board as normal to their allotted paddocks, and there it is.'

Though tough in business the old companies were jealous of their reputation for fair dealing. In 1933 Euston Young wrote to the newly appointed manager of the AP Company's *Yerranbah* outlining the dilemma between ethics and commercial prudence: 'Be particularly careful as to how you deal with buyers. Any chance remark from you may create an undesirable impression in his mind. In what I say below I am not suggesting at all that there should be anything whatever in the nature of deceit. The Company would expect that the most scrupulous honesty be observed in all our dealings. But the following position for instance might arise. Under exactly similar circumstances and with perfect honesty you might express the opinion that the sheep are only fair, the country not too good and prospects bad, when I might say that the sheep are fairly good, the country not too bad and the prospects reasonable. We might both be equally honest. But when I, as general manager, express the above opinion, and am told that your opinion is quite different the position for me is difficult, and my only escape may be to say that you are a fool.'

The wool crisis of the early 1970s caused a major restructure of land ownership, and many famous names passed from the scene. The big old English pastoral companies, which had done so much to settle the inland, lead herd and flock improvement and train young people, were looking decidedly shaky. Most of them were broken up and had left the stage or were sold to local organisations by the end of the 1970s, victims of government closer settlement policies, high levels of taxation, low returns, the impatience of investors, and in some cases asset plays. Pastoral house Dalgetys, which had purchased Pastoral Development Holdings and its string of properties from the Killen family in the late 1960s, acquired that venerable giant New Zealand and Australian Land Company with its 20 large and historic stations, and immediately began to sell off. First to go was *Dromore*, of 6175 hectares just 14 kilometres from

Cooma. Others were to follow, including such famous properties as *Bundure* and *The Yanco* near Jerilderie and *Wingadee* at Coonamble. The English-owned Australian Estates was sold to sugar-miller CSR, which was to eventually sell off the properties piecemeal. The Scottish Australian became embroiled in a complicated and rather ugly round of takeovers involving the Darling family, a firm called Philip Macquarie, and New Zealand-based Industrial Equity. Soon they had sold three of their larger properties – *Rockwood* at Hughenden, *Bowen Downs* at Aramac and *Uabba* at Lake Cargelligo. The rest were gone within a few years. In many cases, the local Australian companies and institutions who acquired these holdings, and who should have known better, got their timing wrong and sold out at the absolute bottom of markets during the 1970s and 1990s. The confidence of woolgrowers was shaken when the Squatting Investment Company, which for nearly 90 years had been primarily a wool producer, turned itself into a cattle operation. At their annual general meeting in April 1972 Chairman JS Balderstone announced that in 1971 more than 81 per cent of their revenue was from cattle, and that they had shifted out of wool because 'we have no faith in an industry which must rely on subsidy'. Three of the company's sheep properties had been sold the previous year – *Mt Manara* at Ivanhoe, *Quantambone* near Brewarrina, and *Kynuna* in north-west Queensland.

Some companies had attempted to ride out the first wave of closer settlement in the 1920s by 'sacrificing' properties; splitting them into small blocks for sale to settlers on extended terms. The Union Mortgage and Agency Company of Australia, later to become Australian Estates, had taken possession of *Brookong*, near Lockhart in New South Wales, in 1895 from William Halliday. At the time it was 85,600 hectares shearing 220,679 sheep and lambs, producing 2602 bales of wool, and the debt on the property was an incredible £284,428. During the 1900s they began selling blocks on extremely generous terms. By 1919 they had reduced the holding to just under 28,000 hectares, which were gradually sold down until the final blocks went in 1926. In 1922 Norman Clapperton, then inspector of stations for the company, reported: 'The evidence of our intentions, as far as possible, to meet any legitimate demands for closer settlement will, we hope, have the effect of diverting the New South Wales Government's more or less compulsory settlement proposals to other landholders who have failed in this respect.'[14]Most of the old shareholder-based pastoral companies did a great job in running their stations in a controlled and ethical manner. They had developed a deep understanding of the cyclic nature of the business, and looked after their country by stocking it lightly. When they disappeared there were few organisations left that provided a career path for young men and women. Investments in the pastoral industry are best judged over 20- to 30-year spans, which expose the business to a range of market and seasonal cycles. By the 1990s few investors had the patience for such a long exposure or the volatility of returns; they tended to measure their investment returns over yearly or even quarterly time spans. So after the old companies had gone, few organisations ever again attempted to acquire and run big strings of sheep properties that complemented one another by being spread over a range of regions and climates, or set up such detailed reporting and data-gathering systems.

There are still station managers running large properties for various family and syndicate investors. They operate in a more complicated and tougher environment that requires a mix of strategic and tactical skills. The practical issues and the challenges of seasons and markets still exist, but much more information is available about them and a manager needs the skills to access and analyse masses of it. Modern society demands that they be less autocratic and more consultative in style. At F.S.Falkiners during the 1980s and 1990s, when the company was owned by the media group News Limited, three of the local management team sat on the board of directors. Monthly meetings were held with the seven managers to cover both policy and operational issues, and on-site staff sat on a series of committees handling various aspects of the business. At *Portland Downs* in central Queensland in 1988 the 'autocratic management style of Cecil Luck is a thing of the past. All employees participate in monthly staff meetings. These meetings are not mere formalities, as management action results from the consensus achieved at them.'[15]

The station manager of 2000 is faced with many of the same dilemmas that perplexed earlier generations. In some ways the dilemmas are more acute, because their new city-based property owners often lack the industry wisdom accumulated by the old companies or understanding of seasonal and market cycles, and their expectations of regular returns on investment may be too high. Peter Forrest described the classic dilemma: 'The manager's chief difficulty was to apply in practice the sometimes rigid and theoretical dictates of the owners. Seasonal and other conditions could change extremely rapidly, and a manager tied down to inflexible orders could soon find himself in a very difficult position: "Obey your orders and you will break your owners" was the commonplace manager's maxim.'[16]

CHAPTER ELEVEN

SOLDIER SETTLERS

Ten pounds of flour, 10 pounds of beef, some sugar and some tea,
That's all they give to a hungry man, until the seventh day.
If you don't be mighty sparin', you'll go with a hungry gut –
For that's one of the great misfortunes in an old bark hut.

ANON, *A SWAGMAN'S SONG*

In my lifetime, and my father's, one strong thread rewove the social fabric of the bush more than any other thing – government-sponsored schemes to split up large properties into smaller blocks and distribute them by ballot. The motives behind closer settlement were laudable – to increase national productivity and create a stalwart yeoman class with its roots in the land. The idea was great, but the execution was disastrous.

Australia's dry interior had been opened up illegally. In the early and middle 19th century a breed of tough and restless entrepreneurs known as squatters moved inland, grabbed huge tracts of frontier land beyond the fringes of settlement, and built up flocks of sheep numbering in the hundreds of thousands. Historian Paul McGuire described the era: 'The squatters came, driving their beasts out beyond the borders of settlement and squatting on the land. In the 1820s the government had forbidden settlement beyond prescribed areas under its control. But as one chastened governor remarked, it would be as easy to confine an Arab to a circle drawn in the desert sand. Out in the trackless (and lawless) bush they drove, building bark huts in the wilderness, fighting drought, floods and devastating fires. Some died quickly from thirst and native spears, but the flocks and the wool grew, and the government could see that from the squatter was coming the first great exportable commodity. By the 1840s the squatters were growing wealthy. By the 1850s they were wool kings in command of the country. The squatters became an Australian aristocracy, and those first hard men, who fought the wilderness and fought the law, were the stuff from which primitive aristocracies usually appear. They are the great figures of Australia's barbaric age, lean, hard, shaggy men who often could neither read nor write, who sent their heirs to Oxford and Cambridge, and took their titles and outstared the sun.'

Amid the bounding post-war optimism of the 1920s a new group of pioneers was to appear, equally tough and persistent, who were to suffer trials and deprivations every bit as hard as those of the 19th century, but mostly without their ultimate reward of affluence. These were the soldier settlers. When troops returned to Australia during 1919, the government faced up to the issue of how to rehabilitate them into civilian life. There was no counselling available in those times, nor would it have been welcome in the prevailing culture of robust self-sufficiency, but there was a dim recognition of the psychological adjustments they needed to make, and concern about reports of drunkenness and brawling among restless returned men in Melbourne.

A succession of governments had enacted legislation to 'bust up the big runs' and 'squeeze the squatters', starting with the *Closer Settlement Act* of 1904, which provided that large runs could be resumed, the owners compensated, and the land made available for small farms. It was an attractive proposition politically, and in 1919 was reinforced by the patriotic notion of settling returned heroes on a block of land. Prime Minister WM Hughes put legislation in place as early as 1917 that provided Commonwealth financial support for the states in a comprehensive soldier settlement plan.

State governments wasted no time. In 1920 the *Trawalla Estate*, 17,000 hectares in western Victoria, *Gringegalgona*, Balmoral, of 12,000 hectares, and *Mt Bute* at Linton of 19,000 hectares were acquired by the Victorian Closer Settlement Board. By 1925 the board had purchased 3350 properties comprising 700,000 hectares for £13.2 million. Demand for blocks was strong in Queensland. When four selections from *Westlands* at Longreach were thrown open in 1921 there were 789 applicants. In 1922 three blocks from *Bimerah*, Isisford, attracted 993 applications. That year 2157 selections totalling 2.3 million hectares were taken up in

Queensland. In 1923 the government gave notice to resume another three stations, and in 1927 more than two million hectares were made available in the Longreach district alone.

The squatters' runs were certainly being busted up. In the 32 years from 1892 to 1924 the number of properties in Australia running 50,000 sheep or more had shrunk from 214 to 34. The number running from 1000 to 10,000 sheep had grown from 4455 to 7386, and those running from 100 to 1000 from 8317 to 13,565. This trend towards fragmentation was later to prove dangerous, particularly in the arid lands.

There were interesting experiments in sponsored settlements. In 1922 outrage was expressed at the suggestion that retired Indian Army Officers be settled in the Victorian Mallee country. 'Indian Officers, as workers, will be of no use whatever, they are used to having everything done for them, and if they have enough to retire on, they may settle down in the sea-coast towns and help swell the big cities'.[1] But the project went ahead, and in 1923 a type of Kibbutz was formed at *Leslie Manor* near Ballarat, with 23 Pukka Sahibs in residence at the homestead. Australian Farms Limited was to manage the property for three years, while the officers did their share of the work and gained experience. After three years they appointed their own manager and ran the place on the group system.

It was richly symbolic that many of the colourful pioneers were passing from the scene in the early 1920s just as the big runs were being cut up. They included the astute Robert James Martin McBride, who had arrived in 1856 with five shillings in his pocket and walked from Adelaide to Burra. There he took on mining, fencing and carting, and in three years acquired 518 square kilometres of leasehold country, then another 1295 square kilometres towards Broken Hill, then *Faraway Hill* and *Outalpa* and *Outrina* stations, all in the back country. There was also the charismatic Cuthbert Featherstonehaugh, defrocked Church of England minister, fearless horseman, fighter of bushrangers, manager of *Brookong* when it carried 90,000 sheep, and later owner of *Goorianawa* and *Mungerie* in the Coonamble district. Robert Gray, who had fought in the Indian mutiny, came out in 1863, overlanded to the Hughenden district in north Queensland and took up 407 square kilometres of virgin country. The resilient John Waugh, out from Scotland in 1854, worked at *Boonoke* when it was a cattle run, later managed *Burrabogie* and *The Yanco*, and took up *Clare* at Balranald, where his descendants are to this day.

The whole structure of soldier settlement was doomed to failure – blocks were too small and their owners were under-capitalised. By as early as 1923 a total of 913 selections had been forfeited or abandoned in Queensland. A Soldier Settlers' Defence Association was formed in Victoria to protect their rights in the political arena – at the time overdue payments to the Closer Settlement Board stood at three million pounds. Settlers were demanding that the size of holdings in the Western Districts be lifted from 120 hectares to 200 hectares. In 1927 the Victorian Government revealed that since 1917 it had advanced £30 million on the scheme, and that £23.55 million of this was still outstanding.

In Queensland things were little better. In 1917 around 600 square kilometres had been resumed from *Wellshot*, and divided into blocks of 2000 hectares. This was a political decision contrary to the advice of the government's own Land Commissioner, who recommended blocks of 8000 hectares. Because of the insignificant size of the holdings, the financial stress put upon the 27 selectors, and the bitter squabbles that broke out about boundaries and water, the troublesome settlement became known locally as 'The Balkan States'. More far-sighted was a commission of inquiry appointed by the Queensland Government in 1927. They recommended that in the central regions of Longreach and Blackall a 'living area' should be sufficient country to run 6000 sheep, in the Winton/Hughenden area 8000, and in the Jundah/Windorah area 10,000. It is a significant comment on the inability of governments to learn from history that 20 years later, following the Second World War, a 'living area' in Western New South Wales was considered to be an area sufficient to run only 3000 sheep. This was a misjudgement that was to prove disastrous for the welfare of the settlers and of the natural environment.

In 1938, when the Victorian Government finally scrapped its closer settlement policy, more than half the soldier settlers who had been placed on blocks – 6677 men and their families – had walked off or had been forced off. Many ended up destitute on the streets of Melbourne during the Great Depression of the 1930s. Remaining settlers and their families endured lives of bitter toil. Poor housing, inadequate food, and hard, bullocking, physical work from daylight to dark was their lot. In current terms all lived below the poverty line, but the optimistic hope of building a better future sustained them for a while, until circumstances just beat them down.

Among those who survived was Alfred Bush, who was discharged from the army in 1919, and in 1921 drew 119 hectares of black soil near

Rapunyup in the Victorian Wimmera. In 1999 his son David recalled: 'The block Dad got was completely virgin. He paid six pounds an acre and had 45 years to pay it off at interest between one per cent and two per cent. Life was hard; for the first three or four years he lived in a tent made out of old hessian bags, spending what money he had on second-hand machinery and livestock.'[2] Bill Forrest had drawn 4000 hectares off *Tara* in the Longreach district during 1920: 'At first I just camped out or stayed at *Caithness*. Later I built a hut near the *Willowie* bore, and this was home for me and whoever I had working for me. Things were pretty rough, and there was no running water or refrigeration or refinements like that. We shared the cooking and lived mostly on meat and damper. Our meat was almost all salted and put in bags and then hung up in a tree to keep.'[3]

Everyone was short of cash, and it was the settlers' wives who made sure the family was self-sufficient, scrimping and saving, making and mending clothes, cutting the family's hair, growing vegetables, raising chooks and feeding pigs. They endured lives of constant drudgery, cooking over wood stoves or open fires, laundering clothes in primitive conditions and baking bread and damper. Isolation was a real issue for those not used to the back blocks. Often without a telephone, the settlers' wives felt it keenly. When their husbands left for work, usually early in the morning, they had no adult contact for the rest of the day, and this often went on for week after week. Few families could afford a trip to town more frequently than once every few months.

Loneliness and isolation affects people different ways. When selling rams during the 1960s I used to visit woolgrowers in far south-west Queensland, where properties were a day's drive apart over rough and unmade tracks. Some of these people had telephones, some were linked to the outpost radio, and some had neither. Some welcomed people from the outside world, and if visitors failed to stay two days and drink a bottle of rum each night, they would be deeply offended. Others had fallen in love with isolation; children would hide behind the furniture, the wife would hide in the kitchen, and the husband would only speak in strangled monotones without looking up. They could hardly wait to be rid of you.

Some few soldier settlers of the period after the First World War bravely survived and prospered. The majority did not. The government's estimation of what constituted a living area was far too optimistic, so the scheme caused personal misery and deprivation, and was also to have serious long-term environmental and social effects. The settlers' struggle to survive often meant that land was overworked, so it contributed to some of the environmental problems that still haunt inland Australia. The old fiercely independent entrepreneurial bush culture was eroded – people had been encouraged to take up blocks by governments; they had suffered sorely and felt betrayed. Many took the view – with some justification – that governments put them there, so governments should bail them out. For their part, governments – with some collective sense of guilt – were inclined to offer various forms of support over the next 60 years, and in the process inhibited the economic rationalisation that would otherwise have taken place. It was tragic for Australia's economic development and for another generation of young Australians that succeeding governments were to make the same mistakes following the Second World War, with the result that the wool industry was to barely survive the tough economic times of the 1970s and 1990s.

Why did brave and worthy people take up a challenge that had so little chance of success? Few people today would tolerate the discomfort, hard physical toil and lack of leisure endured by the settlers. Their commitment was partly in fulfilment of the Australian dream of standing independent and self-sufficient on your own block of land, partly a symptom of the driving optimism of the post-war world, partly a desire to emulate the pioneers and found a dynasty. As Peter Forrest put it: 'In every instance the selections were places of striving; where people battled to make a living from the land and struggled to create an inheritance.'[4]

Despite clear evidence that the scheme was in trouble, federal and state governments continued to acquire big properties for subdivision during the 1930s, giving preference to returned soldiers. In 1939 more than 75 per cent of growers were producing clips of fewer than 30 bales – in reality a clip of 300 to 400 bales was needed to ensure long-term viability. The industry was becoming overcapitalised as each settler built a homestead, a machinery shed and a woolshed, and bought a tractor or a team of horses.

The Victorian scheme was in the deepest trouble of all. In 1930 a deputation from the British Overseas Settlers' Association called for an inquiry, claiming misrepresentation, maladministration, lack of practical supervision, provision of unsuitable stock, worn out implements and clapped out horses – all of which had combined to bring the settlers to an intolerable position. A Farmers' Relief Bill was passed to give protection from creditors – this bill established a board that was to control and apportion farmers' income amongst creditors. By 1932 more than 50 per cent of the 27,939 settlers in Victoria were receiving protection or assistance, the majority were technically insolvent, and the commission had outstanding liabilities of £34 million, secured against the doubtful asset of £25.4 million due in payments from the settlers. Yet the madness went on. In the mid-1930s the government announced the opening of 10,000 hectares in the Suggan Buggan district of East Gippsland, to be split into 800-hectare blocks suitable for running 500 sheep.

The blocks being offered in Queensland during the 1930s were generally larger and the land better suited to Merino sheep, so demand remained strong. In 1935 three blocks taken from *Terrick Terrick* and *Lorne* at Blackall attracted 1358 applicants. There were 867 applicants for five blocks taken from *Thurlagoona* at Cunnamulla.

In 1936 the New South Wales State Cabinet noted continued demand for blocks, but that little crown leasehold remained. They called on the owners of large freehold estates to subdivide their holdings as a national duty, but, in case patriotism was found wanting, stipulated that if a satisfactory response was not forthcoming there would be compulsory acquisition of sufficient properties to meet demand. Yet in 1938 special legislation had to be passed to save hundreds of small farmers in the Lake Cargelligo and Hillston districts from ruin through impossible levels of debt.

Managers of company-owned stations were often applicants for blocks. Company jobs, though modestly paid, carried with them a career path, reasonable security, a comfortable lifestyle, social standing and power. So what drove them, of all people, to take on undeveloped blocks in a harsh environment and at great risk? Perhaps Jill Ker Conway provides the answer. Her father left a station manager's job to take up a 12,000-hectare block near Mossgiel in south-western New South Wales, and she wrote of her parents: 'Neither of them could explain to curious children where their urge to get ahead and make something of life came from. Different as their backgrounds were, both were driven to excel, to run the best station, breed the best sheep, raise the brightest children, run the most efficient house.'[5]

If anyone deserved to prosper it was the post-First World War settlers – that is if courage, persistence, hard work and just sheer guts count for anything. Against all the odds, some did, and their doing so was a tribute to the fine qualities of bush people. It is a relief to read Jill Ker Conway

A soldier settler's humble home at Barmera, SA, in 1922 *(SLSA: B 62210, State Library of South Australia)*.

writing of how her family began to prosper as wool prices improved from the mid 1930s: 'After three years of such effort, the annual wool sale produced a surplus. It was possible to make a trip to Sydney for dental and medical care, and the luxury of buying some stylish and well-cut clothes. My father could also afford help for the property. Beside the house there now appeared a cluster of outbuildings. Stables and a cowshed, a garage, and a cottage for the station hand. Housing for drays, farm carts, and a lightweight sulky used for fast travel over country roads when rain had made them impassable for cars. My impending birth had also added a telephone line.'[6] At last they were gaining improvements to their lifestyle. 'Each year after the wool sales some new comfort was added. A kerosene refrigerator appeared to replace the old burlap drip safe, bringing iced water into our lives. My mother acquired a 1934 Ford V8 sedan, with leather upholstery and unbelievably comfortable springs. Our new clothes came from mail-order catalogues, and an occasional case of crisp red apples would arrive.'[7]

As the Second World War drew to a close, further subdivision of large properties for closer settlement – giving preference to returned men – was clearly going to be a political imperative. As early as 1942 New South Wales Minister for Lands JM Tully flagged his government's intention to promote closer settlement through the medium of heavy taxation on large estates. Land tax would exempt properties with unimproved capital value below £5000, impose a light tax on properties valued up to £20,000, but exceeding that it 'would be gradually increased until it was at a rate that would compel subdivision'.

The debacle associated with post-First World War closer settlement was fresh in people's minds, and in May 1942 a warning was sounded by *The Pastoral Review* in an editorial headed 'Will the old mistakes be repeated?': 'Closer settlement as carried out provincially has been a national calamity. The political catch-cry of closer settlement in the mouths of cheap-jack provincial politicians has been responsible for huge increases in state debts, calamitous losses and, worse still, has resulted in great damage to the lands of Australia. Repatriation of soldiers, or land settlement for any purpose, after this war is as likely to result in bungling, scandals, waste, colossal debt and damage to Australian lands. Already by wrong use of our soils we have created dust bowls, and we have areas where the plough has proved to be more deadly than the rabbit.'

In 1944 Federal Cabinet approved a set of principles under which a closer settlement scheme would operate. It was to be a joint venture between the Commonwealth and states, and the principles went some way towards recognising past mistakes. They provided that settlement would only be undertaken in areas where economic prospects for the settlers were sound, and the number settled would be determined by the opportunities rather than the number of applicants. Qualifications and experience of applicants were to be assessed, they were to be allocated sufficient land to enable them to earn a reasonable labour income, and technical advice was to be available free of charge. Lack of capital was not to preclude suitable applicants, and this was to be achieved through interest rate concessions. All advances made during the first five years were to be interest free, during the next five years at two per cent, and thereafter at market rates. It was hoped that this would give settlers 30 per cent equity in their venture after 10 years. It all sounded terrific.

Prime Minister Curtin announced in 1944 that, to obtain sufficient land for settlement, legislation would be passed to give the government powers for compulsory acquisition, and all rural land prices were to be controlled. The *War Service Land Settlement Act* provided that the price to be paid for land in New South Wales, Queensland and Victoria was to be based on 1942 values. At the same time owners of city property acquired by the Crown were compensated at ruling market values. This inequity was to be the subject of some nasty litigation during the 1950s.

Generally a 'living area' in the pastoral zone was deemed to be sufficient country to run around 3000 sheep. This was to prove adequate for the booming markets of the late 1940s and early 1950s but completely inadequate when commodity prices adjusted back to long-term trends in the 1960s. As one commentator said in 1945: 'The basis of a living area always seems to be worked out on the assumption that every year will be a good one seasonally and on the further assumption that the products of the land are always going to be readily saleable at the highest prices.'[8]

February 1945 saw moves in Victoria to acquire 34 pastoral properties in the Western Districts with a total area of 109,000 hectares. They included *Woodhouse*, 5000 hectares at Penshurst and *Langi Kal Kal*, and 10,000 hectares at Beaufort. In May that year New South Wales Lands Minister Tully announced that 6300 potential settlers had been interviewed, and inspection and proclamation of 113 estates totalling 800,000 hectares were underway. These included *Sunny Ridge* of 5100 hectares near Blayney, *Berida* at Coonamble of 16,700 hectares, *Waterloo*, Inverell, of 5500 hectares, and *Wantabadgery East* and *Wantabadgery West* near Wagga of 14,000 hectares.

In Queensland the first post-war ballot was held on June 26, 1946 in Hughenden. Blocks of 8000 to 12,000 hectares were from *Cambridge Downs*, *Maxwelton* and *Manuka*. February 1947 saw the opening up of blocks in the Western Division of New South Wales, including four 12,000-hectare blocks from *Cultowa*, Wilcannia, and parts of *Warraweena* at Bourke. The largest single acquisition at the time was *Edgeroi* near Narrabri, where 48,000 hectares of rich black soil were split into 70 farms ranging from 370 to 1000 hectares. In January 1949 the Commonwealth Government announced they had so far approved 2.7 million hectares as suitable for ex-servicemen, allocated 1.8 million hectares to 2213 settlers, and expended £6.5 million on the scheme.

There were personal tragedies along the way. Charles Leslie (Les) Falkiner was the eldest son of the family patriarch Otway, and, though talented, his personal development had been stunted from living in the shadow of a charismatic father. A returned serviceman, he was allowed to participate in three land ballots, eventually drawing one of the blocks from *Wantabadgery West* near Wagga. A political storm then brewed up over a member of a wealthy grazing family being allowed to hold a soldier's block, so eight happy and productive months after taking possession a special Act of Parliament was passed to deprive him of ownership. A thwarted and wasted talent, he returned to the protection of the family company, where he slowly slipped into inactivity and alcoholism.

Often the original station owners retained some part of the property, and their attitude to the new neighbours varied. Some studiously ignored them, and apart from the normal civilities acted as if they were not there – 'Let them stew in their own juice,' as one grazier put it. Another north-west New South Wales grazing family, noted for their thrift, made the settlers lives as uncomfortable as possible. If water supplies were shared, they cut them off. Before handing blocks over they put in teams of axemen to cut all fencing timber available and stacked it inside their new boundary – knowing full well that under terms of their leases settlers were required to erect certain yards and fences, so would have to buy the material back. They made sure that blocks were stripped bare of pastures and anything that was moveable; what was not moveable they made so with an oxy-torch. Others were like Angus Waugh at *Clare*, north of Balranald. *Clare* was 181,000 hectares at the turn of the century, and 81,000 hectares were still retained by the family going into the Second World War. The Closer Settlement Board intended to take the lot, and it was only after Waugh convinced them he was a First World War Digger that they left him with the 12,000-hectare homestead block. A tough brown gnome of a man, who played hard and worked hard, Waugh was wise in the ways of sheep and the vagaries of arid land. He was generous to the settlers in offering that wisdom, helped with the supply of rams, and guided them through the tough early years. For nearly a generation they would drop in at the old homestead to hear his advice and views.

During my time as a jackaroo in the 1950s, managers of the large remaining stations would often invite neighbouring settlers over to Christmas parties or the movies that the stations usually ran every few weeks. The strain of the battle that these people were fighting was etched in their faces; we could sense their shyness, their awe at the machinery, the facilities, and the living conditions that people on the stations enjoyed compared to what they were enduring. Even we jackaroos, the humblest of employees being paid a pittance each week, felt embarrassed for them, with their broken-down old cars, shabby clothes and air of defeat.

Deterioration in woolgrower's terms of trade steadied the momentum of closer settlement, though there was great concern in 1951 when the New South Wales Government proclaimed for future resumption four famous Riverina Merino studs, *Deniliquin Stud Park*, *Wanganella*, *Zara* and *Wanganella Estate*. The Stud Sheep Breeders' Association and the New South Wales Graziers' Association wielded the considerable political clout needed to have the proclamations lifted. As demand for blocks diminished preferential treatment for returned soldiers in land ballots was no longer applied. In 1952, 10 blocks in the Walgett-Brewarrina area – taken from *Gundablui*, *Dunumbral*, *Angledool*, *Milroy*, *Dumble* and *Charlton* – three at Bourke from *Toorale* and *Corella*, and two at Hillston from *Manfred* were all open to civilians. In Queensland the extensive leasehold properties of the Australian Pastoral Company in the Dirranbandi area – including *Noondoo* and *Cubbie* stations – were being split up, despite the company offering their *Amby Downs* freehold to the government in exchange. Josiah W Spence – a 74-year-old widower of *Mundah*, Dirranbandi, where they bred them tough – drew a *Cubbie* block of 5000 hectares in the ballot. He declared he was in excellent good health and the prospect of starting hard work on his new property did not concern him; he was anxious to get on with it.

Originally, land for settlement was acquired by the state governments at 1942 values, and though this had been relaxed to the extent of adding 15 per cent, offers were still below ruling market values. Station owners were pinned to the wall, because if they refused the offer, then land was compulsorily resumed anyway, at the base 1942 value. In effect, long-term landholders were being forced to subsidise closer settlement. The Pye family, who had a reputation for toughness and business acumen, and with extensive grazing interests in north-western New South Wales, vowed they were not going to give in to blackmail. Their *Ghoolendaadi* property, of 15,200 hectares near Boggabri, was proclaimed for resumption in September 1950 at a valuation of £209,549. The Pyes claimed the true value was £544,800, and appealed to the Land Valuation Court in April 1952, where Justice Sugarman sympathised but found against the family. They then appealed to the State Full Court and won, with costs awarded against the government. Not to be beaten and with a lot at stake, the government sought leave to appeal to a higher court and threatened to enact retrospective legislation should they lose in that arena. Eventually the case went all the way through to the High Court of Australia, which a year later found against the Pyes. Many landholders received a nasty surprise when the court pronounced that a state government had the constitutional right to confiscate the property of any citizen without any recompense whatever and need not, as the Commonwealth Government must, provide just terms of compensation. Grimly determined, the Pyes took the case to the Privy Council in England, where they again lost. Eventually the New South Wales Government increased their offer by £32,541 to £242,000, and in 1956 *Ghoolendaadi* was split into 30 farms for ballot.

Prompted by a looming election rather than any twinges of conscience about their behaviour, the New South Wales Government relented, and in November 1955 agreed to pay fair market value in future for land resumed for closer settlement which, of course, had largely been completed. A sad reflection on the whole closer settlement scheme as it applied to returned soldiers came in 1959 when New South Wales Minister for Lands Gollan released a report showing that 1607 soldier settlers still owed nearly a million pounds in overdue payments, while 456 had forfeited their holdings for non-payment.

In retrospect, soldier settlement following the Second World War avoided some of the mistakes of the 1920s. But it did put too much pressure on the land as people overstocked and overfarmed to survive, and it did create production units too small to cope with the low commodity prices that were to come with the 1970s and 1990s. Like their fathers 25 years before, few of the post-Second World War soldier settlers fulfilled the dream of founding a pastoral dynasty. The blocks were too small and organic growth too difficult. But their sons and daughters were the baby boomers and, conscious of their own battles to make ends meet on the land, parents tended to invest in education. Many more bush children completed secondary education and went on to university than had done so in the past. Generation change and the reluctance of these better educated children to return to the farm saw large numbers of original soldiers' blocks put on the market during the troubled 1990s. Fortunately, booming land prices in 2003 and 2004 enabled many to get out at reasonable values, and properties began to aggregate into bigger areas again. Those few families who stayed on their original holdings survived by compromising lifestyle, living self-sufficiently, taking no holidays, enjoying few luxuries, with few trips to town, and family members taking what off-farm work they could get.

At the start of the 21st century land was slowly being returned to bigger operating units, purchased by neighbours or city-based investors. The land-buying onslaught from public companies and institutions fishing the bottom of the market, about which many farmers had held irrational fears, had not come to pass. City buyers tended to be high-worth family partnerships or family companies who had made money from the new industries, or had collected a monumental severance package, or sold their company. They had always wanted to own a farm, but not just a hobby farm – they sought and found properties of sufficient size, or put blocks together, to build a commercially viable unit upon which they could exercise their commercial skills.

It seems likely that long-term viability in the arid pastoral lands requires no less than 25,000 sheep per operating unit. The clock is being turned back 50 years, and it needs to be. Wool is an industrial commodity, despite the prattle of many wool commentators who claim it is not and lament the alleged deleterious effects of a supposed 'commodity culture' among growers. It is in the nature of commodities that they fluctuate in price, and their value decreases relentlessly in real terms over time. So the keys to running a business in a commodity industry are managing market risk and reducing production costs per unit. Continually reducing costs requires the ability to increase scale of production. From the 1970s onwards downstream users of wool – the topmakers, spinners, weavers, makers-up and retailers – have been aggregating into bigger and more powerful units with enormous commercial leverage in the marketplace. Early-stage industrial processors in particular have embarked on a restless search to relocate their activities into developing countries with low labour costs – first into Japan, then Korea, Taiwan, and eventually China, India, Mexico and Eastern Europe.

Thanks to the closer settlement policies of a succession of governments since the 1920s, wool production in Australia remains fragmented, with many thousands of small producers. A large proportion of the clip is being produced in the wheat/sheep zone, where flocks are integrated into the farming rotation, and wool is virtually a by-product. Generally individual growers have little ability to lower costs through increasing scale of production, hold little market power, and are often weak sellers. Commercial leadership is lacking because there are few large producers with power in the market – the top 10 to 15 growers in Australia together probably produce not much more than one per cent of the clip.

If left alone, the production end of the wool industry would, over time, have rationalised itself without too much social dislocation, as successful managers gradually absorbed the less efficient – just as happens with restaurants and corner stores, where the process of rationalisation is constant. But, because of the iconic status of wool growing, generations of politicians, fearful of electoral backlash in the inland regions, have been unable to restrain themselves from interfering. Subsidies, drought relief, interest rate support, 'special circumstances' payments, taxation laws, land tenure laws – all have played their part in ensuring that wool growing remains a fragmented industry. If the vast stations that existed in the period up until 1920, many running flocks numbering in the hundreds of thousands, had not been split up into quite so many small parts by well-intentioned closer settlement schemes, the industry would have entered the 21st century in better shape.

CHAPTER TWELVE

RABBITS, RABBITS, RABBITS

A small, rather likeable-looking animal, possessed of no great ferocity, vegetarian in habit and shy by nature, was to alter the course of Australian history. Because of the rabbit, Acts of Parliament were made, personal fortunes were gained or lost and the Australian environment was adversely affected by the onslaught of tens of millions of small chisel-like teeth and sharp nails.

BRIAN COMAN, *TOOTH AND NAIL*

The rabbit scourge reached its height at the start of the 1950s, and was then broken by the myxomatosis virus. At their peak rabbits were costing the Australian pastoral and agricultural industries $600 million a year in lost production alone – livestock-related productivity lifted more than 25 per cent after the virus had done its work. The damage they did to the environment was incalculable, and is still evident today. As Brian Coman said in *Tooth and Nail*, his authoritative history of the rabbit in Australia: 'We tend to think of the bush being destroyed by the ringing axe, wildfire, the chainsaw and bulldozer, but rabbits can achieve the same results – it is just that the time scale differs.' Rabbits and closer settlement were an evil combination.

The sheer number of rabbits infesting the country at their peak in 1950 is difficult to describe to those who did not witness it. 'The paddock was literally a moving mass of rabbits' is a cliché, but it was true. At sunup and sundown, when rabbits are most active, the landscape just swarmed with a

Rabbit extermination drive in the 1950s *(The Age, Fairfax Photos)*.

dense grey cloud of the obnoxious host, which was demolishing pastures, destroying whole species of vegetation, ringbarking trees, creating dust bowls, and soiling everything in sight. Those of us who worked the stations, who saw and understood the damage rabbits were doing but felt powerless to do anything about it, grew to hate the little animals. We found them repulsive.

Thomas Austin of *Barwon Park* near Geelong, whose family were pioneers of the Merino sheep industry, is infamous for introducing rabbits – for hunting – to Australia in 1859. He probably bears too much of the blame; Governor Phillip carried rabbits as cargo in the First Fleet, the Henty family brought rabbits to Portland in Victoria in 1834, and they were reported near Castlemaine in 1845. Whoever was to blame, they rapidly spread north, and by 1879 were as far inland as Wilcannia in north-western New South Wales. In 1881 runs were being abandoned in the Mallee and Wimmera areas of Victoria because of rabbits; that year they were also reported around Balranald and 200 kilometres to the north. They reached southern Queensland by 1886, and were found at St George and north of Cunnamulla in 1890. Coman attributes the speed at which they spread to three factors – absence of natural predators, soils suitable for burrowing, and cropping of pasture by sheep, which afforded rabbits the short grazing on which they thrived.

As infestation turned to plague, and plague to devastation, desperate property owners and governments sought the weapons to fight an all-out war. The first was a huge system of rabbit-proof netting fences, and between 1885 and 1914 more than 300,000 kilometres were erected, including a fence stretching from Corowa in southern New South Wales to Barringun, north of Bourke. A variety of poisons were used, including phosphorous, arsenic and strychnine. Deadly substances like strychnine in one or two ounce tins, calcium cyanide in five-pound tins, and chloropicrin in two-pound bottles could be bought over the counter. In those days the use of protective clothing was considered a waste of time, a practice for the effete, and Coman's father told him that after 'a day's work on the phosphorous cart you could rub your hands together and they would glow in the dark.' Every manager had his favourite bait to entice rabbits onto the poison trail. They included apples, pollard, cereal grains, chaff, jam, thistle roots, carrots and dried quinces. My father studiously avoided two varieties of jam – apricot because that was all they had to spread on their dry army biscuits during the long months on Gallipoli, and quince because he had used tons of it for killing rabbits.

In 1887 the New South Wales Government offered a reward of £25,000 to anyone who could find a way to control rabbits. They received responses from all over the world – including one from the French microbiologist Louis Pasteur – which ranged from the fanciful to the lunatic. They included using diseases such as smallpox, syphilis, glanders, hydrophobia and tuberculosis. There were even proposals for electrocution. Attempts were made at biological control when some local infection killed rabbits on *Tintinallogy* on the Murray River. One suggestion, given seriously, was to feed dogs on the infected carcases, dose them with a purgative, and allow them to roam over rabbit-infested areas. As Coman says: 'It required faith in the method to ride out with 200 purging dogs.' Another imaginative proposal came from William Rodier of *Tambua* at Cobar. He suggested turning the promiscuity of buck rabbits into a weapon by concentrating on killing does only, so that the bucks would end up tearing one another to pieces in fights over the remaining does. Populating infested areas with cats was another idea, and in 1886 four hundred cats arrived by rail at Bourke, headed for *Tongo* station near Wanaaring. They apparently soon arrived at a truce with the local rabbits.[1]

All sorts of gases were pumped down rabbit burrows: exhaust fumes from vehicles, sulphur, charcoal gas, various arsenical mixtures, and 'Cynogas', which was calcium cyanide dust. The poison gas used against troops during the First World War was tried, despite warnings from veterans who had personal exposure to it. Coman described a 'fearsome machine, very much like a flamethrower, which generated the deadly phosgene gas. A kerosene-fed flamethrower provided sufficient heat to convert carbon tetrachloride, in a small dripper tank, into phosgene, and the whole deadly concoction of flames, combustion productions and phosgene was blasted down the warren.'[2]

Most of the bigger stations employed two or three rabbiters around the run, living alone in unlined tin huts. Solitude had turned many of them into eccentric 'hatters'; I remember one chap who talked to himself incessantly, and another who spent his spare time teaching his dogs to yodel in chorus. They fought a losing battle of attrition using dogs, traps, ferrets, poison and fumigants. They were paid a modest wage and made pocket money from the sale of skins. It was a great sight to see the rabbit packs working when I was a boy at *Buckiinguy*. Truckloads of assorted dogs would come up from the dog's home in Sydney. They were all shapes and sizes – mastiffs, kangaroo dogs, greyhounds, miniature terriers, bitzers, beagles and bassets. When rabbits were put up in open country the whole pack was off in an instant, howling, barking, yelping and squealing. When the rabbit was caught or went down a burrow, dogs would strut around and scratch the ground in that self-important way that dogs have, as pleased as punch with themselves, even the bassets and beagles who arrived at the scene long after the action was over.

Rabbit drives were common during the late 1940s and early 1950s. These were carried out at dawn or dusk. After securing the fences, a line of beaters walked across the paddock, carrying wooden waddies and driving all before them. Eventually the host was driven into a netting cage in the corner of the paddock – a great writhing mass of hatred perhaps waist deep. Then each of those thousands would be taken out and have its neck wrung by grasping it behind the head and by the hind legs and stretching it across the knee. The soft click as the neck broke lives in my memories still.

The use of Ferguson tractors for ripping burrows after the Second World War gave the property owners a slight edge, but by and large rabbits survived everything that was set after them, pumped at them or trailed out for them. Then came salvation in the form of a viral disease called myxomatosis, which was mostly spread by mosquitoes. The disease was first identified in Uruguay in 1896. Cultures were sent to Australia in 1919, but nothing was done because of politics and the importance of the export trade in rabbit skins used for making felt hats. There were further experiments in the 1930s, but these were officially discouraged because rabbits ('underground mutton') were a cheap alternative source of protein during the hungry depression years. Then the virologist Dame Jean MacNamara, whose advocacy had prompted resumption of work during the 1930s, revived the concept of using the virus, but her work was held up because of wartime restrictions. In 1950 a trial began near Gunbower on the Victorian side of the Murray. In early 1951 the disease broke out spontaneously and spread like wildfire. By the end of January infections were reported as far north as the Lachlan and Murrumbidgee rivers, and by June it had spread across much of southern Australia.[3]

Results of the disease were not pretty to see. Roads and surrounding paddocks were littered with dead and dying rabbits, emaciated, blind, their eyes swollen and gummed shut with pus, their fur coming out in clumps. But the old hands like my father, who had spent the best part of a lifetime fighting a bitter war against rabbits and had developed a deep hatred for the pest, felt few pangs of pity or remorse. On visits to Sydney, when some thoughtless hostess served up a nice baked rabbit, he would leave the room in disgust.

Myxomatosis broke the back of the rabbit problem in Australia. The population has waxed and waned since, with genetic resistance to it and other viruses, but provided property owners remain vigilant rabbits will never again devastate the Australian landscape. Make no mistake, rabbits did more damage to the environment than European farming methods, overuse of water, salt, land clearing and overstocking put together. Had rabbits still been in plague numbers during the extended drought of the early 2000s the devastation would have been beyond contemplation.

An industry and the environment had been saved. Coman tells of an interview with the conservative western Victorian grazier Arthur Beggs: 'Arthur Beggs started to show me photos and statistics of what it had been like before the disease struck and then suddenly he broke down and started to cry. Tears just poured out of him, but he said "Don't worry these are just tears of relief. Myxo has saved us".[4] Dame Jean MacNamara, who died in 1968 and is buried at Beechworth in Victoria, is a true Australian heroine. Her statue should stand in every wool-growing town and in every conservationist's office.

CHAPTER THIRTEEN

DROUGHT TAKES HOLD

It's grand to be a Western Man,
With shovel in your hand,
To dig your little homestead out
From underneath the sand.

AB Paterson, *It's Grand*

Sing a song of saltbush,
Sandy-blight and drought,
Forty thousand weaners
Slowly pegging out.
Start 'em out on the stock route,
Brand 'em with a 'T',
Isn't that a pretty sight
To show a mortgagee?

Anon

Drought is a grim business. The first sheep to die are those in what is known as 'the tail' – the five to 10 per cent of any mob that have poor constitutions or are carrying some disability. Then the rest gradually become weaker and weaker. First their hipbones begin to stand out like hat racks and they walk with heads drooping; then the ribs begin to show. Many will contract 'night blindness', a condition caused by a vitamin deficiency, and blunder about running into trees and fences. Starving ewes quickly lose their milk supply and maternal instincts, so walk away and leave their lambs to die. Then they develop a distinctive short-stepping, tottering gait, which indicates their end is not far off. Many will die round the edges of tanks, too weak to escape the mud after they take a drink. Any attempt at sudden movement will cause sheep to fall, and once down they will never rise again. The crows begin to circle, stalking round and round until they gain sufficient courage to dart in and take the eyes out. Soon the winds are picking the wool from pitiful little piles of bones that litter bare paddocks. It is a stark battlefield.

The only two absolute certainties about living in outback Australia are excessive heat and periodic drought. My generation was brought up on the story of the seven-year-old boy who had never experienced rain in his life. One day it did rain, and he got such a fright that he fainted. They had to throw a bucket of sand over him to bring him round. In my case the first part of the story was nearly true, for I was nine before I saw really heavy rain and a bountiful season around our home in central-western New South Wales. My working career of nearly 50 years included coping with five serious droughts. My father lived through the two most destructive droughts of the 20th century, and as a child during the first of those suffered the trauma of seeing his family have to walk off their 280,000-hectare *Florida* station in the Cobar district.

In January 1920 drought and a heatwave hit western Queensland: 'It may be mentioned that at Winton the thermometer has been up to 116 degrees Fahrenheit, and in some parts of the town more, for days. As many houses in western towns are merely galvanised iron, the fortitude of the people in such weather must be imagined. Birds dropped dead from trees, fowls died, and in some cases dogs went mad, and sundry persons collapsed. The railway tracks buckled in places. One traveller records that when he stopped the birds fluttered round his water bag, and as he poured water into a pannikin they actually alighted on his hands.'[1]

The greatest of all droughts was between 1898 and 1902, and the only one that has yet approached it in destructive capacity was that of 1944–45. Statistically, the drought of 2000–04 may have been the worst of all, but people were better able to cope because of improved transport, available feed supplies, and various levels of government support. Above all, the rabbits had gone. Consequently, stock losses were not so devastating.

My father saw drought visit New South Wales in the early 1920s. In May 1920, a western grazier wrote: 'It is two and a half years today since we had beneficial rain, and I have been cutting scrub for nearly 20 months for the stock. The situation is beyond drought, it is now absolute famine. There is little hope of saving any of our sheep.'[2] Ninety per cent of the people in his district had spent up to their bank limit. Banks had not foreclosed on mortgages, but had taken over management of the situation, and a bank representative was travelling from place to place controlling and supervising expenditure. The giant New Zealand and Australian Land Company reported in January 1920 that mortality had reduced their total sheep flock by 140,000 to 1,341,328. At Longreach the Australian Pastoral Company reported carrying capacity of their *Warbreccan* station had fallen from 127,420 sheep in 1916 to 45,000 in 1920. In June 1920 New South Wales Premier John Storey, unable to find a political solution, sought divine intervention, and requested the churches of the state to appoint a day of prayer for rain. To everyone's surprise, magnificent drought-breaking rain fell within days. Later in the year drought losses in New South Wales were estimated at a staggering 8.5 million sheep, 423,000 cattle and 74,000 horses.

In the late 1920s the cycle came round again, particularly in Queensland. In 1927 Andrew Williamson, chairman of the Australian Estates Company, reported that on one of their Queensland stations 62,000 sheep were shorn in 1925 and 30,000 in 1926, and fleece weights had fallen from eight pounds eight ounces per head to four pounds one ounce per head. Total losses from all their stations were 140,000 sheep and 11,000 cattle. Statewide, sheep losses were nearly four million by the end of 1926. The Queensland Government was unable to organise divine intervention, but eventually rains came and restocking began. In September 1927 a consignment of 3000 ewes for one owner arrived at Townsville from Sydney via the steamer *Tarcoola*, with another 9000-plus 200 rams to come.

Tactical management of drought was different in those days when Australia produced little grain compared to the huge areas sown to crops from the 1960s onwards. Grain for feeding stock was scarce and expensive, so when bad seasons closed in graziers sought to lease or purchase relief country in areas where there had been rain. In 1920, after the New South Wales state premier had organised intervention, Otway Falkiner reported the return of 27,000 ewes to their Deniliquin properties from the hill country round Tallangatta on the Murray and at Camperdown in western Victoria. But by 1927 they were again seeking agistment for 40,000 sheep, and inspected the Yarram Estate in western Victoria, then 9300 hectares, with a view to purchase at four pounds 10 shillings per acre. In 1926 the owners of *Maneroo*, at Longreach, walked 30,000 sheep south to find grass at Hebel on the New South Wales border. That year there were 250,000 sheep on agistment in the Emerald district, 120,000 around Normanton, and 500,000 in the Gulf Country.

Closer settlement and the splitting up of large holdings meant that not only were more people being adversely affected by droughts, but the wool industry as a whole was less resilient to their damaging consequences. Chairman of the Australian Pastoral Company Keating took a shot at Australian politicians and their closer settlement policies when he spoke out at their London meeting in February 1928. 'Surely the prejudice of the more advanced Labourites [sic] against pastoral lessees, who they describe as men amassing wealth by monopolising vast areas to the exclusion of poor settlers, should not apply to companies with thousands of shareholders, many of moderate means, whose capital is developing land, and in the possession of stations so separated, that it is unlikely that all would be subject to drought at the same time.'

Fifteen years later the unwelcome visitor called again. As if being at war was not bad enough, during the 1940s Australia suffered the most destructive and protracted drought since that of 1898–1902, which is still recognised as the worst of all. Encouraged by profitable wool prices locked in through the wartime wool acquisition scheme that began in 1939, growers had increased their sheep numbers. The national flock, which stood at 111 million at the outbreak of war, had grown to 126 million by 1942, the highest since 1894. Then drought began to slash numbers as sheep starved and died, so that by 1947 the flock was reduced to 95.7 million – a tragic loss of more than 30 million animals. Individual districts suffered horrific losses. In October 1944 it was estimated that in Queensland around Dirranbandi, Bollon and St George 40 per cent of the sheep had been lost. In 1945 Hunter Patterson, president of the Riverina Graziers' Association, reported that properties in the central Riverina were carrying only 40 per cent of normal sheep numbers, and the western Riverina was down to 20 per cent. In the Balranald district, where there had been 535,000 in 1940, numbers had shrunk to 75,000. In 1947, after the drought had broken in Queensland, the South West Graziers' Association stated that their members were depasturing 500,000 sheep compared to a million in January 1943. Losses in 1946 alone had been

30 per cent, and in the Maranoa district were as high as 70 per cent of sheep and 50 per cent of cattle – many had lost the whole of their flocks after a futile effort to save them.

Woolgrowers adopted fairly standard strategies for combating drought. Initially they had to make a decision as to whether what they were experiencing was a real drought that might persist for a year or more, or just a dry pinch that might be over within a few weeks or a few months. Their decision could only be based on a study of rainfall records, experience and sheer gut instinct; the chances of getting it right were slim. If it was assumed that drought had set in then non-core stock, such as wethers or old and inferior sheep, were sold to give valuable breeding ewes priority over remaining pastures. When pastures were completely eaten out then came the agonising decision whether to market core stock quickly before they became unsaleable, handfeed, or seek agistment country elsewhere. By this time others would be flooding sheep on to the market, so prices would be low, and there was the certainty that when drought broke replacement sheep would be scarce and expensive.

Restocking to restore the income base after drought posed a huge problem. Most woolgrowers came out of drought in a debilitated state financially, having borrowed to handfeed stock and pay fixed overheads, and now like Oliver had to go back and ask for more. Other problems could intervene – in 1949 the Winton district in Queensland was carrying only 300,000 sheep compared to a normal population of 1.125 million. An attempt to purchase 40,000 restocker sheep had to be abandoned because the Liquid Fuel Board refused the application for petrol to deliver them by road.

Preserving a flock through the agistment or hand-feeding options meant borrowing funds or drawing on reserves. It would be certain that as drought deepened and widened, demand for feed grain would increase and prices would go through the roof. There was the ever-present risk that if the crisis dragged on funds could run out or grain become unprocurable, so sheep could die anyway, after a substantial investment in keeping them alive. Overlaid on the physical demands of feeding stock was an emotional dimension. Most sheepmen felt great affection for their stock, and had made a substantial commitment over years to lifting their genetic quality through classing, selection, and purchase of expensive rams. The sight of distressed and dying animals, dragged out over many months, brought on feelings of guilt, depression, and eventually despair. Many took each death as a personal blow, and blamed themselves for the suffering. Drought was a tough proposition both financially and emotionally.

The most depressing feature of any drought is to run out of household water. The homestead at *Buckiinguy* drew its water from the Macquarie River, which in the 1940s stopped flowing downstream of Warren, and eventually degenerated into a series of small turbid pools stinking with dead fish. We were carting domestic water about 16 kilometres from a government tank, using hand pumps and a Furphy water cart. Rigorous water discipline was imposed in the household. No-one ever washed their hands under a running tap; we all used a tin dish with a splash of water in the bottom. At night we took turns bathing in a few centimetres of muddy water. Once I was admonished severely for taking a drink from the water bag and thoughtlessly tipping out the dregs from the glass.

When the drought eventually broke I can remember walking with my father along the dry bed of the river to see fresh water coming down, fully expecting to have to jump for the banks when confronted with a rushing wall of water. So I was somewhat disappointed when we at last came to a miserable trickle slowly creeping into the cracks.

Dreadful dust storms were a regular occurrence. The rabbit plague was near its peak, and even before drought set in they had destroyed much of the surface vegetation, so topsoil began to move, driven by howling westerly winds. These storms were frightening in their intensity. They moved in slowly from the western horizon until a dark, threatening mass loomed overhead, often thousands of metres high. They were usually preceded by an eerie stillness such as that before a thunderstorm, and during that lull there were frantic preparations. Homesteads were sealed tight, curtains pulled shut, lights turned on, and people moved inside to cower in the choking heat. When the storms broke it was as if nature had gone mad. The atmosphere howled and moaned, debris flew around like shrapnel, darkness closed in and visibility came down to a few metres. They could last for hours or for days. In 1944 it was reported from Longreach: 'A severe "Bedourie Fog" came in from the south-west on October 16 and lasted for three days with an intensity seldom seen in these phenomena. On the first two days the limit of vision was about half a mile, and all planes were grounded.'[3] The aftermath was squalid. Dust intruded everywhere and surfaces were smeared with a film of it. There was grit between your teeth, in your ears, in your eyes, everything you touched grated. Any pastures left standing were sandblasted and rendered unpalatable to stock.

The constantly shifting sands, carried by westerly winds, began to pile up against north–south fences, and by 1946 many were completely covered so that stock could simply walk over them. When rains eventually came these banks set into a solid mass. The journalist Lionel Krone visited *Uardry* in the Hay district in 1946: 'The north–south boundary fences of *Uardry* attracted our attention. Wooden stakes had been erected to take another two lines of wire, for only one original wire was showing above the ground, the well-grassed earth sloping up from the western side. Out on the back of the run many of the fences were showing only a few inches above the sand.'[4]

The course of the 1940s drought can be tracked through the fortunes of the Falkiner family at *Boonoke*.[5] The company turned in a modest profit of £26,000 in 1939–40, but things were looking grim, as Chairman Otway Falkiner reported to his board: 'In view of continued drought I have closed for the lease of *The Burra*, 19,500 acres four miles from Tumbarumba, for an annual rental of £7750. I plan to send up 20,000 sheep from their

The oil painting *Murrumbidgee Whaler* changed hands from the Falkiners to the Fields in a 1945 deal. It still hangs in the Sydney office of T.A.Field Estates *(Photo Mike Bell)*.

Riverina properties. I have already sold as many surplus sheep and cattle as possible, but the remainder are in very poor condition and unsaleable. Other sheep will have to be handfed at home.'

In January 1941 he reported the Riverina country was in bad shape, but stock were holding out on supplementary rations, which were costing £400 a month. He anticipated that unless good rain fell in a matter of weeks handfeeding would have to be resorted to on a large scale. Later that year the Union Bank was putting the pressure on and advised that 'should its overdraft limit be exceeded, as it appeared certain to be, no increased accommodation will be given, and it will be necessary to sell something, possibly land, or raise further money by mortgaging freehold land.' The wish for rain was granted, and there were phenomenal falls in January, but it was only teasing, and winter rains failed. Things became progressively worse.

In 1944 Otway described the country as 'reduced to the condition of a desert'. The November 1944 board minutes noted: 'Mr OR Falkiner reported that he had inspected the various properties and had to report that the pastures were shockingly bare despite the drastic reduction of sheep and cattle numbers.' He was given leave to buy whatever feed was necessary. By January 1945 there had been no relief, and the feed bill for the year was £30,000. There had still been no rain by March of that year, and the bank overdraft was £48,210. A recent inspection of the stations and their stock indicated the situation was desperate. As the only agistment country on offer was far away in the Moree district, it was decided to feed all remaining sheep at home.

Otway Falkiner, like most big pastoralists of his time, was a decisive man and autocratic in his management style. But beset by difficulties, pressed by the banks, and still stricken by the loss of his favourite son in the war, even his formidable energies were becoming exhausted. Near the end of his tether, he began to include his board and his sons and nephews in the critical decisions that had to be made.

By the end of May 1945 the overdraft had gone out to £61,812, and fodder supplies in eastern Australia were running out. Otway had adroitly secured supplies of lucerne hay in the Hunter Valley, but at the last moment the government stepped in and cancelled the sale, because they claimed the hay should be supplied to milking cows. This was a bitter blow, and in later years Otway claimed the hay had in fact gone to racehorses. He then managed to secure 9000 bags of maize in the Glen

Innes district at eight shillings and six pence a bushel, a colossal price. He reported to the board: 'The price is high but the position is desperate.' When delivery of the maize was slow, he obtained a considerable quantity of vegetables out of the Sydney markets, mostly pumpkins and potatoes, but these proved useless.

It was at this time Falkiners were forced to sell their entire drop of maiden stud ewes – a very serious step of last resort, because these represented the genetic future of the Merino studs and would be generators of future revenue streams. The buyer was the astute TA Field, whose family owned a string of pastoral stations in New South Wales and Queensland and who was also a ram customer of Falkiners. The drought broke in August 1945 before the ewes were delivered, and Otway phoned with an offer to buy them back, knowing that Field had him cornered. Field, a long-sighted man, realised the quality of his future ram supply was at stake, so agreed to sell the ewes back to Falkiner at the same price – cancelling the deal. But only on condition that Otway gave him the oil painting of a 'Murrumbidgee Whaler' (swagman) that hung in the smoking room at *Boonoke North*. It hangs to this day in T.A. Field Estate's Sydney office.

Of course when drought breaks it does not rain grass, and generally it takes a fortnight for pastures to germinate, a month to develop a green pick, and six to eight weeks for feed to grow. After a prolonged drought like 1940–45 the seed bank is depleted so recovery is slow and patchy. Then restocking has to be financed. By August 1945 the Falkiner overdraft had ballooned to £82,214. The Union Bank had reluctantly agreed to a temporary increase of their drawing limits, but by February 1946 the overdraft was up to £107,272, and the bank was restive. Otway was sent for by McMullin of the Union Bank, who advised that, contrary to earlier agreements, he was in receipt of a cable from his London directors ordering him to give no additional overdraft exceeding £113,000, that first mortgage security was required, and the company must sell land. Falkiners immediately opened confidential negotiations with the National Bank and Dalgety and Co., which was prepared to lend £30,000 for purchase of stock, secured against the stud sheep flocks.

At a board meeting in February 1946 it was decided Otway should seek an interview with the Prime Minister and Treasurer, to 'put the position to them in the hope that in view of the importance of our industry to Australia some drought relief should be granted to us or, at least, Land

Tax remission.' Otway reported in August 1946 on his interview with Chifley and Scullin: 'I was promised consideration under provisions of the Act. Compensation was refused and the Government representatives would not agree that our business was of any national importance – the interview was disappointing to me.' Conservative graziers and Labor federal governments were not a good mix.

Otway Falkiner later recalled those times with great bitterness: 'The government had commandeered all the transport that would have brought feed to save the dying stock. We battled to get feed. It cost us £98,000 to save such of our sheep as we could. Even then we lost 10,000 of them.' Many of the losses were actually suffered when drought-breaking rain fell. Dust carried in the fleeces set into solid blocks, and in their weakened state, with their digestive systems disturbed by the sudden availability of succulent green feed, sheep went down and never got up.

The Falkiner family had fought a gallant fight against overwhelming odds. The injuries inflicted on the company by the 1940–45 drought, during which they accumulated trading losses of more than £150,000, were not fatal, because the industry was entering the period of greatest prosperity in its history. But it did leave the company severely maimed and ill prepared to fight the next serious drought 20 years later.

Old man drought ravaged the country again during the middle 1960s – just 20 years after his previous devastating visit. The pastoral country was stricken during the two years from 1964 to 1966, and wheat/sheep areas in 1965 and 1966. The general atmosphere of fatigue and depression caused by drought increased the bitterness of the debates on wool promotion funding and the reserve price scheme, which were being fought out at that time. Whole families were out in peeling caravans, walking their flocks around stock routes chasing grass; children were taken out of school and taught through correspondence by harassed mothers. Once drought set in it seemed as if it would never rain again, and flock owners poured over weather maps and rainfall statistics to find any glimmer of hope. With wool prices falling, many once-proud woolgrowers sold off their flocks and restocked with cattle when rains came.

At the height of the drought sheep could hardly be given away. In April 1965 a pen of good Merino ewes in the Riverina of New South Wales made only two shillings a head. In May of that year 535 sheep railed from Nyngan to the Sydney market sold for four shillings and seven pence a head. The bill for rail freight came to eight shillings per head.

Drought and lower sheep numbers decimated the trade in Merino rams. Flock ram sales dwindled away to nothing, and the 1968 Sydney ram sales saw the worst market in 40 years; only half the rams offered were actually sold. In February 1966 federal and state parliamentarians attended a crisis meeting in the town of Cobar in western New South Wales. Distressed graziers reported that many had lost half their sheep. A recent survey showed that 170 graziers in the Bourke district had lost 276,000 sheep between them, and their total drought fodder bill had reached $376,000. In the New England region, usually a kinder climate, one station in the Guyra district reported losses of 14,170 sheep, 376 cattle and 32 horses. In February 1967 it was reported that in Queensland 5.6 million sheep had died since the start of the drought, nearly 25 per cent of the state's total sheep population. In some of the northern areas graziers had lost more than a third of their stock in the 1965–66 season.

Many graziers, and local storekeepers who remained unpaid, had to cut their losses and sell up. At that time I was a professional sheep classer based in the New South Wales town of Dubbo, and my business just evaporated – no-one had any sheep left to class. A client in the Gilgandra district owed me two years' fees and, unable to pay, gave me 600 wethers instead. I found a paddock and cut scrub to feed them for eight weeks, and their sale paid the grocery bill. The Falkiner family took a bold decision when feed cut out on their Riverina properties; between February 28 and March 10, 1966 they moved 30,000 stud sheep to agistment in the Walgett–Brewarrina districts of New South Wales where decent rain had fallen, using more than 100 trucks. Falkiner employees were stationed on each of the agistment properties, including *Dunumbral*, which carried 10,000 head, and one of the directors, TK Falkiner, took up residence in Walgett to supervise operations. But things had changed since the days when Otway Falkiner shifted his flocks around in times of drought; the freight bill alone came to more than $350,000, and the exercise was one of the factors that finally crippled the F.S.Falkiner family empire, forcing the shareholders to sell four years later.

The 1960s drought was certainly a disaster for many worthy woolgrowers and their families; it also shaped up to being a national calamity. In August 1965 it was announced Australia's foreign reserves had fallen by £71 million or 8 per cent, largely as a result of drought and low wool prices. And worse was expected. Various authorities were making anxious comparisons with earlier drought episodes. They noted that during the 1898–1902 drought 47 million sheep had been lost, more than half the national flock, and it took 32 years to recover the numbers. In the 1940s drought 29 million were lost, and it took 11 years to recover the numbers. But the 1960s industry was vastly different. Despite predictions of doom, and tragic losses incurred by some individuals in some states, losses from the national flock, which stood at 170.3 million in 1965, were limited to 12.5 million, or 7.3 per cent. Numbers were restored within three years, and within four years they were at an all-time high of nearly 180 million.

Many things had changed. Over the preceding 20 years woolgrowers had been assiduous in improving their properties, dividing them into smaller paddocks and putting in more watering points. This meant that sheep could make better use of available pastures. A higher proportion of the national flock was being run on improved pastures in areas with higher average rainfall where drought, though still serious, tended to be less protracted. Use of motor transport meant stock could be moved to market or relief country faster and with less stress. Australia had also become a major grain grower, so there was access to supplements for hand feeding. Most importantly, managers had become smarter, with better access to information on weather patterns, sheep nutrition and available markets.

Though there were to be further droughts in the 1970s, 1980s, 1990s and again in 2002, never again would the Australian sheep flock be decimated by reason of drought alone. The long drought that broke in 2003 has been called the worst ever experienced, and it could well be in statistical terms of rainfall deficiencies. But assessing the financial and strategic impact of drought is a very difficult business, because of the powerful emotional charges that build up during what amounts to a long, drawn-out crisis. Heart-rending stories appear almost daily in the press, usually accompanied by an appropriately horrific picture of dying animals, or of tearful children carrying sad little suitcases as they walk out the farm gate with their parents. The competition for news being what it is, these stories are usually selected from the worst-case category. During the course of a drought there are also those, mainly the woolgrower representative organisations, with a vested interest in talking up the devastation, and those, usually government bodies with the task of allocating various forms of financial assistance, with a vested interest in talking it down. Drawing the line between what amounts to a national disaster, hence worthy of support by the taxpayer, and what is standard business risk that should be factored into the price of land is extremely difficult, and it shall always remain blurred and uncertain. So drought ends up as a political issue – usually the worst of all arenas for the making of rational decisions.

The 2002–03 drought had one very significant difference to its predecessors – it was the first in my lifetime during which sheep prices remained high and there was a ready market for surplus animals. We can be assured, however, that had rabbits still been around, had things been the same in terms of transport and the availability of grain as they had been in 1902, then things would have been much, much worse.

CHAPTER FOURTEEN

THE GREAT DYNASTIES CRUMBLE

My name is Ozymandias, King of Kings:
Look on my works, ye Mighty, and despair!
Nothing before remains. Round the decay
Of that colossal wreck, boundless and bare
The lone and level sands stretch far away.

PERCY BYSSHE SHELLEY (1792–1822), *OZYMANDIAS*, 1819

One day in 1951 the ageing Otway Falkiner had a visitor. It was an old companion from his roaring days, 'Black Alec' McPherson, who had once supervised pastoral interests for the Chaffey family along the lower Darling River, and whose son Bill owned a neighbouring property. Otway was 77, his sight failing and his once formidable physical and mental powers ebbing away. As Bill McPherson drove them around the 8900 hectares of *Boonoke North*, which had been Otway's home for most of his working life, the two old friends chatted in the back. They came across a mob of sheep. 'I think these are the top rams, but I'm not sure – can't see them properly,' Otway grunted. Then after a pause: 'You know Alec, I'm buggered; not up to it any more. But my family's buggered too – they are not up to anything.'

It was a turning point. Otway Falkiner led the second generation of a great pastoral dynasty founded by his father Franc Sadleir Falkiner in 1878, and as chairman and chief executive of the family company he controlled a huge land and livestock empire in the Riverina. He was facing up, in his pragmatic way, to classic third-generation problems. Within two years he was to take the brave and then unprecedented step of appointing an outsider, the brilliant visionary Basil Murray Clapham, to run the family company. Clapham, at the time Victorian manager for wool brokers A.M.L.&F., was originally approached by Otway to restore the quality and reputation of the Falkiner Merino studs, which had deteriorated as the old patriarch lost his grip. After a preliminary visit Clapham said to his old friend: 'Boss, the sheep are not the problem, it is the family. The only way I can do it is to have full control over the company and its affairs.' So Otway bit the bullet, and in 1953 Clapham was appointed general manager of F.S.Falkiner and Son Pty Ltd, with a brief to restore the company's fortunes, on the then princely salary of £6000 plus purchase of 2000 hectares of land under generous interest-free terms over 10 years.

Otway Falkiner was one of the very few heads of the great pastoral dynasties with the courage to make such a move. In the period from the early 1960s to the mid-1970s many of these dynasties, which had done so much to develop the pastoral industry, were to crumble and be blown away by the fierce winds of change.

Otway's father Franc Sadleir Falkiner was one of the eager and ambitious young men who migrated to Australia in the 1850s with modest capital, attracted by the vibrant economic activity created as a result of the gold rushes. He joined many younger sons of Scottish and Irish families who were to found dynasties that made a tremendous contribution to the commercial, pastoral and cultural development of the new nation. Falkiner, like others of his ilk, was entrepreneurial by nature, and soon went on a frenetic search for commercial opportunities. Suzanne Falkiner tells of the early flowering of Franc's commercial instincts as a child in Ireland: 'On an occasion when all Daniel and Rebecca's sons were given a guinea to spend as they wished, Daniel, the eldest, went out and bought a flute. Franc, however, promptly bought a supply of potatoes, planted them, and sold the crop at a profit.'[1]

On arrival in Australia Franc travelled directly to the goldfields, where he kept a store and bought gold for the banks, making enough to repay his mother £20 she had lent him when he left home. He married in 1856, moved to Ararat, opened a grain store and bought 1600 hectares of land in the 1870s, then shifted to Geelong in 1875 and operated as a land tax

The Falkiner family in 1939 (left-right) top row: Bill, Mac, Travers, CLS (Les), Fraser, John and Leigh (jnr) Falkiner; middle row: Frederick Knight, Ralph, Otway and Leigh Falkiner, and John Carse; bottom row: Franc Falkiner *(F.S.Falkiner archives)*.

valuer. Then he set his eyes on opportunities developing in the pastoral industry, and in 1878 with two partners purchased *Boonoke*, in the Deniliquin district of New South Wales, from the Peppin family, whose unique and productive strain of Merino sheep was about to become the most sought-after in the land. The price was £60,000 on a 'walk-in–walk-out' basis, and he grew his empire from there. He bought his partners out in 1882, then came the purchases of *Moonbria* in 1884, *Tuppal* in 1891, *Moira* and *Morago* in 1899, *Pericoota* in 1900, and *Warriston* and *Widgiewa* in the early 1900s. He put his five sons to work managing the properties, paid no dividends, and constantly marshalled his capital to fund expansion of his flocks and herds. Land was relatively cheap at the time, and livestock expensive, so he seldom sold female animals, and as soon as he bred up sufficient numbers he bought another property. It was a colossal effort, and his grandson Fred Knight captured the times: 'Today it may seem strange that one man, in the space of not much more that 30 years, could create such an empire. They would think it strange because they have lived in an age when taxation has been so severe that it is almost impossible for a man to become wealthy through his own honest earnings. It was not then impossible for a man with ability, energy and good judgement to accumulate enough capital from savings and shrewd investments to embark on an enterprise which, from modest beginnings, would bring him a fortune.'[2]

Falkiner and the other emerging pastoralists of his era were strong personalities, highly individualistic, and robust both mentally and physically. They were decisive in their ways, affecting an abrupt and sometimes rough manner, spoke their minds freely, often had booming voices, and tended to shout at people. Falkiner still had plenty of rough edges to him when he bought *Boonoke*. Apparently he stayed overnight at *Wanganella* on his way to inspect the property, and the women of the Peppin household were deeply shocked by his habit of repeatedly spitting the quid of his chewing tobacco into the fire, where it hissed and bubbled. These men ran their empires with an iron hand, consulted few people when making decisions, and were autocratic in their ways. Fred Knight wrote of his uncle Ralph, one of Franc's sons, who invented the first electric shearing machines and mechanical sugar cane harvester: 'Many people said he should have been an engineer, but he had no formal training. He may have wanted to be that, or he may not. Whatever his real wishes, Grandfather would never have consented. He had created an empire for his sons and it was their duty to manage the properties in it. None of his sons opposed him during his life; his word was law.'[3]

Franc Falkiner and his peers were pastoral entrepreneurs, and followed closely behind the squatter pioneers who had opened up the back country for livestock. They brought in new capital and new ideas, and consolidated their empires. They had little prior experience with the pastoral industry or of breeding and managing livestock, but made shrewd judgements about people and employed good men. Falkiner acted quickly to put on that prince of sheep classers JC Darke, leading man of his day, to control breeding of the stud sheep at *Boonoke*. Falkiner's generation remained unsentimental and objective about their businesses, and would quickly dispose of any asset, person or animal not paying its way. They usually eschewed working with their hands, and delegated the hard physical toil to others. The second generation, on the other hand, were brought up in the business and almost from childhood learned to use a pair of shears, mark lambs, muster paddocks and understand the intricacies of day-to-day hands-on management. They grew to love the business for its own sake.

They developed a deep commitment to their specific industry and became steeped in its lore and traditions. If asked, Franc Falkiner would have defined himself as a 'businessman' or 'pastoralist'; the second generation like Otway would have responded 'stud sheep breeder'.

Most of the second generation were diligent. Otway was physically strong, with a stocky frame and restless energy. Probably the greatest ram salesman of all time, he constantly bustled around the country seeking new customers. If it rained in western Queensland he was soon away to Sydney by train, to Brisbane or Rockhampton by ship, then inland by car to visit the customers. Then he would appear back at *Boonoke* going round the stock, or bob up at *Moonbria* or *Zara* checking pastures. He enjoyed robust discussion, and could seldom be bested in an argument. He once arrived at *Boonoke* in a flurry of activity and instructed manager Billy McKees to drive him around the run. Spotting a section of new fence, he expostulated: 'What a stupid place to put a bloody fence! What silly bastard decided to build it there?' McKees answered quietly, 'You did boss, on your last visit.' Otway paused and took another tack: 'Money, money, money. All you blokes can think about is spending money.'

In 1955, by which time his eyesight was poor and he had become forgetful, I used to read Otway's mail to him each day in the *Boonoke North* office. One day he had just lunched with the Hay district woolgrower and author HB Austin, who was on a visit inspecting rams. As I was reading to him he saw Austin entering the office, and asked 'Who's that bugger?' I replied 'That's Mr HB Austin, boss.'

'No, no, no, he went mad years ago and they locked him up.'

'Well it's Mr Austin all right; you have just had lunch with him.'

'Well, well, then they must have let the bugger out again.'

Everyone was in fear of Otway's wrath. His son OMcL ('Mac') Falkiner, when managing *Boonoke* in the late 1940s, had a great weakness for new machinery and gadgets, and there was a constant procession of salesmen out from Deniliquin to take advantage of it. So frightened was 'Mac' of his father's reaction that prior to his visits of inspection he would have all new purchases hidden away in the dish of a nearby sand hill. Many of these items, still in delivery wrappings, lay there rusting away for the next 30 years.

Otway and others of the second generation inherited their fathers' instinctive, ad-hoc and unstructured decision-making style. They tended to get away with it because of their enormous energy in implementation, and an underlying flexibility. If they decided upon a certain course of action and it didn't work, then they faced up to it and changed direction quickly. This generation, like their fathers', tended to be strong personalities, with a commanding presence and peremptory manner. At ram selling time Otway would organise a bustle of activity in the sheep yards, with perhaps half a dozen of the larger clients selecting their rams at once, and big mobs of sheep moving through the yards. Anyone who got in the way, irrespective of who they were, was dealt with severely. Otway would bellow: 'Get out of the way you bloody blockhead!' The victim, perhaps a customer, would slink away in a huff. A little later Otway would soothe ruffled feathers by ambling over, pinching the victim's elbow in his characteristic way, and saying, 'Sorry for shouting at you old man – but you were in the bloody way!'

These second-generation pastoralists were forceful but seldom arrogant. They had lived through too many hard times, seen their empires too close to collapse as a result of drought or poor markets, to accumulate hubris. Nevertheless they fully expected to be treated in a certain way, and could be high-handed. These days, few would dare write to their bank manager as Otway did in 1923: 'I am in receipt of your letter of 17th inst., and note my overdraft was £14,970. I did not think I was overdrawn so much, but I've been having a rotten time at the races, as I dare say you have noticed. I do not see how it will be reduced for a little while, and I will call upon you when in Sydney next. Meanwhile I shall be obliged if you will kindly give me credit for a further £2000 as I have several accounts to meet.'

Despite extravagant lifestyles, the second-generation pastoralists were unpretentious. The sentiments expressed in Otway's obituary in 1961 were largely true: 'He regarded as an act of distasteful presumption the adoption by anyone of a feeling of superiority over his fellow men.' His generation were powerful men, much courted for their favour and their business, but they hated flattery and importunate behaviour. I once observed Otway seated at a formal dinner next to a clergyman, who was wheedling up to him in a particularly unctuous manner and asked how many sheep he owned. Otway snapped back, 'What the bloody hell is it to you?', and then added to the effect by leaning over and noisily slurping his soup direct from the plate.

His generation was supremely self-confident, and perhaps it was this that gave them an openness of character and a disarming level of candour that could dismay the less robust. It was a surprising trait to find in people born in the Victorian era. Otway had no compunction about telling people uncomfortable truths, or freely handing out advice on everything from station management to how others should manage their sex lives. If he was caught out, or made a mistake of any kind, he would immediately admit his error and move on to deal with how it might be fixed.

This second generation of the great pastoral dynasties seemed larger than life. Towering figures who did as they pleased, they were the stuff from which legends were made. Subsequent generations envied their ability to express themselves through grand gestures, and felt nostalgic about the less complicated and more tolerant times in which they thrived. But the third generation, which lived under their massive shadows, was often stifled and frustrated.

In *The House of Nomura* US author Al Alletzhauser put an extreme view on third-generation dynastic problems: 'There is a saying among Chinese families that it is better to turn the male heir of a third-generation family firm into an opium addict – and get him out of harm's way – than have him risk the family fortune through dissolute spending on women, gambling, or ruinous business ventures.' In regard to the Australian pastoral dynasties, neither symptoms nor remedies were as dramatic as that. However, third-generation problems soon emerged. Third-generation males were better educated than their fathers; most finished secondary schooling at elite private schools in Australia or England, and many went on to obtain university degrees. They were worldlier, more sophisticated, and developed a wider scope of interests. They enjoyed a more comfortable youth than had their fathers and grandfathers, so developed higher expectations of what the world had to offer them, and were more likely to be distracted. Their situation was complicated by the tendency of the second-generation dynastic elders to retain tight control of the businesses into advanced age, and delegation of responsibility was at best cursory. The patriarchs had higher expectations of their sons than they did of other senior employees, and nothing short of exemplary behaviour or performance pleased them. They were not shy in making their feelings known. Otway would berate Les, his eldest son, in front of other employees, and eventually destroyed his self-confidence. Les possessed manifest skills with livestock, but soon lapsed into inactivity and eventually alcoholism.

Others took the Chinese view and preferred their third-generation sons to stay right away from the business. In the years immediately before the Second World War EI Body, master of a family empire based at *Bundemar* in central-western New South Wales, made it clear to his son EM ('Ted') that he could have access to ample money to indulge his whims and weaknesses, but in return he must stay clear of *Bundemar*. Fortunately it didn't work – Ted took full advantage of the high life, had an extraordinary war record, then returned to run the business with great success.

But, dominated by the steely wills and powerful personalities of their fathers, few third-generation males dared to challenge traditional ways and practices. As times grew tougher in the late 1950s and 1960s they tended to look backwards for salvation, rather than take stock of how their world was changing. Their generation was confronted by a new set of realities, new commercial and cultural challenges – the old buccaneering ways were no longer suitable for coping with shrinking markets, rocketing costs and thinner profit margins. The value of old craft skills was eroding as things became more mechanised, more measured, and moved faster. Indeed the business environment of the 1960s and 1970s required a more thoughtful, structured, researched and planned approach. But the third generation saw themselves as men of action like their fathers and grandfathers, and as the English author John Galsworthy noted: 'A man of action forced into a state of thought is unhappy until he can get out of it.'[4] Sadly, many of the third generation had absorbed their father's instinctive and unstructured decision-making habits, but not that generation's fierce commitment, energy, persistence and flexibility.

A classic clash of cultures occurred soon after BM Clapham was appointed to take control of the Falkiner empire in 1953. It was late spring and Clapham – professional manager, workaholic and gifted sheepman – had just finished classifying the top few hundred stud rams, known as 'reserves' and the most valuable sheep in the flock. He was carefully briefing two managers, both members of the Falkiner family, on how he wanted these important animals managed over the coming months. They listened attentively, and then one replied: 'We will get right onto it in a couple of weeks when we come back from the Spring Racing Carnival in Melbourne.' Clapham exploded, and both parties looked at one another in complete astonishment. Clapham could not believe that anyone in his right mind could consider recreation before high-priority work. Equally, the family could not contemplate missing their annual pilgrimage nor believe that anyone would dare suggest they do so. Clapham prevailed, but there was a good deal of sulking and ill humour.

Many of the great family dynasties had become associated with that most iconic of pastoral activities, the breeding of stud Merino rams. Between the late 1950s and 1973 no fewer than six of these families, who controlled parent Merino bloodlines that had dominated the wool production industry since the late 19th century, fell by the wayside. The Austin family had sold their *Wanganella* stud to Falkiners in 1958. The old autocrat Nielson Mills dispersed *Burrabogie* and *Wahwoon* in 1966, as did the Millear family of *Deniliquin Stud Park North.* The other branch of the Mills family was to sell *Uardry* in 1973. The year 1971 saw *Coonong* pass from the hands of the pioneering McCaughey family and, the greatest crash of all, the sale of F.S.Falkiner & Sons Pty Ltd to the Melbourne-based woollen spinners Cleckheaton Limited. The family shareholders had considered selling their company during the early 1960s and had negotiated with both the real estate tycoon LJ Hooker and the opportunistic Latec Investments Company around a figure of two-and-a-half to three million pounds, but the opportunity was let slip. Clapham's strong hand had been lost in 1966 due to family ructions, and in 1967 the company was on the market again for $5 million, with no takers. Finally, at the very depths of the wool depression in 1971 and with shareholders desperate sellers, it went to Melbourne-based woollen spinners Cleckheaton for a paltry $2.6 million. 'We bought it out of petty cash,' arrogantly quipped Cleckheaton managing director FG James, who as a boy had delivered mail to the Falkiners. It was a sad closing to a colourful era.

The reasons behind the collapse of many of the family pastoral dynasties went deeper than classic third-generation problems. Many had never recovered their financial equilibrium following the destructive droughts of the 1940s and 1960s. State and federal government policies were stacked against them – closer settlement, land tax and death duties were all designed to break up the big family estates, and were all too successful in doing so. Taxation laws in respect of tax on undistributed profits precluded family companies from accumulating reserves to be held against tough times. Many of the family companies had ended up with too many shareholders – there were 23 in the body estate when *Bundemar* was sold, 34 in the F.S.Falkiner company. This fragmentation of equity mattered little when the shareholders had been all brought up on the properties and had a sentimental attachment to them, but later generations were more distant and not prepared to tolerate the asset-rich dividend-poor nature of their investment.

When the Falkiner company was incorporated in 1899 shares were issued at £1000 each, and went to male heirs only. The founder Franc Sadleir and his wife received 20 shares each, 'Bertie' 88 shares, Norman and Otway 72 each, Ralph and Leigh 64 each. The five daughters were completely ignored. Tim Hewat noted in *Golden Fleeces*: 'There hangs in the bar of the pub at Conargo a photograph of some piquancy. For it is captioned "The Falkiner Family 1939": it shows 13 men of three generations – but no women.' We can only wonder what additional vigour might have been injected into the third generation if this attitude had been different. It was often said of Otway's daughter Emily that 'she should have been the man'; but she was never given the opportunity to take a management role in the business.

So many of the family dynasties became part of history. Their passing was sad, but industries did not collapse as a result; businesses survived without them, Merino studs were dispersed but their genes lived on. The new entrants who bought them out, such as Cleckheaton and News Limited who followed them as owners of the Falkiner company, injected fresh capital and new ideas, and refurbished depleted assets. Like the founders of the dynasties, they were often unsentimental and commercial in the way they managed the investment. The industry underwent a renewal.

Older woolgrowers still feel nostalgia for the great pastoral families and for the age they lived in. Descendants of the founders still visit their sacred sites, and mourn for what was and what might have been.

CHAPTER FIFTEEN

STUD BUSINESS

The stud sheep business? Half of it is run by the sheep classers, who will sell anything for commission, the other half by the old families, who are nothing but a mob of pig-headed bloody reactionaries.

BM Clapham, Stud Manager, circa 1950.

One January night in the late 1940s, following the annual Goulburn ram sales, Otway Falkiner and Basil Clapham got on the scoot. Clapham was managing the leading Merino sheep stud *Uardry* at the time, and Otway was still firmly in control of the family empire based at Deniliquin, the management of which he was to hand over to Clapham in a few years' time. The two were strong competitors but firm friends, and they held court and talked sheep late into the night in the lounge of the Alceston Hotel, Goulburn's leading hostelry.

Both were returning to the Riverina by train, which went through Goulburn at 2 am, and they gave firm instructions to young Tom Lilburne, then a jackaroo at *Uardry* and later to become an influential sheep classer, to call them in time. Lilburne dutifully went to wake Clapham at 1 am, to be greeted by an angry shout of, 'Piss off you little bastard.' Deciding to try his luck with Otway, he knocked on his door and entered the room, only to have the telephone hurled at him. So, considering discretion to be the better part of valour, Lilburne went back to bed. At 8.30 am Clapham and Falkiner, both locked into the powerful fog of a deep hangover, stumbled into the dining room and promptly ordered bacon and eggs with two double gins each. They then rounded on Lilburne and abused him for allowing them to miss the train. After more gin they decided to call a taxi, and as they bundled into the back seat Falkiner directed the driver to take them to Hay. This came as a shock, because the driver was used to doing nothing more than shuttling his 1936 Ford V8 sedan from the hotels to the railway station; now he was being asked to undertake a 1118-kilometre trip over indifferent roads. Clapham and Falkiner sat the petrified Lilburne between them in the back, and continued to pour good-natured abuse on him the entire long journey home. They turned the trip into a drunken odyssey by stopping for a nip at the various pubs along the way – Gunning, Yass, Jugiong, Gundagai, Wagga, Collingullie, Narrandera, Darlington Point, finishing up at Carrathool. They were not in a pretty state when they arrived home – but both were hard at work the following morning.

These men were captains of the stud sheep industry, which was top of the tree in the world of wool growing. They ran complicated businesses, were deeply committed to their product, worked incredibly hard, often led rackety lives, did as they pleased, and had developed awesome reputations.

It had all begun during the early part of the 19th century. At the beginning were the Macarthurs and their *Camden Park* flock, located just out of Sydney. Writing 50 years later, pioneer sheepman John Ryrie Graham was to say of these sheep: 'From a worthless class of sheep, Mr Macarthur, with consummate skill and judgement, succeeded in producing sheep of the greatest value to commerce. True, he was aided by the finest wool growing climate known, but without the acute perception, judgement, and skill which Mr Macarthur brought to the task, no merely natural advantages could have availed.'[1] As the colony's flocks began to multiply and the foundations of our first great industry emerged, the new pastoralists rushed to buy Camden Park rams in the hope they would reproduce their fine qualities in their own flocks. The business of breeding and selling stud Merino rams in Australia was born.Soon new players were attracted to what was a very profitable trade. This early phase of development saw most of the nation's rams being produced in the 'inside country', which had the benefit of a temperate climate, good shelter timber and reliable rainfall – the Western Districts of Victoria, Tasmania and the Mudgee district of New South Wales. The aim of these early stud breeders was for extreme fineness of wool, and it was their sheep that created the high reputation of the famous 'Botany Bay Wools'. The *Camden Parks* and other early studs grew really superb wool. An early writer described it: 'There was an atoning beauty about the wool that was hard to resist; it flashed with a gem-like lustre: it was so beautifully fine and even, it had such an exquisite downiness of touch, that all other wool seemed base beside it.'[2] Compared to the Australian Merino, which was to emerge in the late 19th and early 20th centuries, these early sheep were small, fine-boned, delicate of constitution, and yielded extremely light fleeces of short-stapled wool. GH Cox of *Burrundulla* at Mudgee, one of the early breeders, recalled in later years that 'he had seen a ram greatly admired that would not cut more than two pounds [around a kilogram] of wool, which was as fine as cobwebs.'[3] A hundred years later the top rams were yielding fleeces of more than 20 kilograms.

With the 1850s and 1860s came a number of developments that were to challenge the supremacy of the old fine-wool types. These included opening up of the arid inland, changes in wool processing technology, a switch in apparel fashion, and an increase in land values.

The squatters on their dry inland runs at first made little attempt to breed rams, and persisted in buying their sires from the fine-wool studs of the 'inside country'. But the inland plain country was vastly different to those temperate areas – the climate was either Mediterranean or subtropical, summers were fierce, droughts frequent, and stock had to walk great distances each day over unsheltered country to find nourishment and water. The delicate constitutions and cobwebby fleeces of the old Merino strains were soon found wanting: 'On the bare plains, near the watering places, an observer must often see large numbers of sheep – perhaps 4000 or 5000 – split up into small mobs, and standing in circles, with their heads thrust under each other's bellies for shelter; clouds of dust continually rising from the stamping of their feet, a scorching hot wind blowing, and a vertical sun pouring its unmitigated heat upon the gasping creatures. This would go near to burn a hole in a brick wall; what wonder then if it destroy the delicate fibre of a fine-woolled fleece.'[4] There was also an economic effect. 'It was a small liability, when land could be purchased for a song, if flocks did not cut over much wool. Later on, of course, when country became fully capitalised, it was essential that large and profitable fleeces should be grown.'[5]

At the same time the British wool market was changing. Up until the middle 1860s woollen manufacturers had been supplying a relatively small, quality market, which required short-stapled fine wool of the 'Botany Bay' type. Then from about 1865 they began to expand into lower-priced woollen goods for which they required longer-stapled wool for blending. At the same time worsted processors needed long-stapled wool to supplement meagre supplies of European product and to compete with French worsted fabrics. So a growing demand in Europe for longer wool coincided with the build-up in sheep numbers in the arid lands of Australia. The stage was set for development of new types of sheep, and the pinch of economic circumstance on a few entrepreneurial pastoralists soon ensured it happened.

The Peppin brothers, George and Fred, were Englishmen who had moved to *Wanganella* in the Deniliquin district of New South Wales in the 1850s from Mansfield in Victoria, bringing their fine-wool sheep with them. Their venture was unsuccessful, so in 1861 they offered *Wanganella* for auction at Scott's Hotel in Melbourne. But there were no bidders. Frustrated, the brothers returned to the Riverina vowing to breed a type of sheep more suitable to the district. Like many sheepmen of the time, they were traders who bought and sold sheep as a side enterprise to their breeding program. They went about selecting the biggest and the best

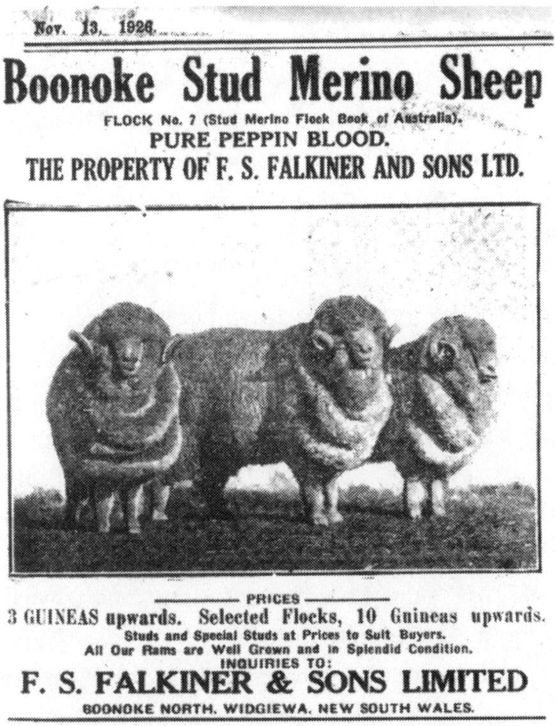

The Falkiners advertise their stud rams in the 1920s *(F.S.Falkiner archives)*.

from the many thousands of sheep that passed through their hands. They imported rams from America and France. They purchased ewes from South Australia, and probably whipped in a dash of the English Lincoln to breed longer wool with natural grease to insulate fleeces against heat and dust. They sought out the renowned pioneer sheep classer Thomas Shaw. Whenever they produced a ram with the qualities they were aiming for they inbred that line to fix the type – their outstanding sire 'Emperor', champion ram at Hay show in 1871, was 'used on his daughters and grand daughters as long as he lived'.

The type of animal that emerged from this process was entirely different to the Merino as it was then known. It had a big, blocky body, strong bones, robust constitution, and could produce a five to six kilogram fleece of 21 to 22 microns with a staple 80 mm long. This compared to a three to four kilogram fleece of 19 microns and 50 mm from the old type. As could be expected, not everyone approved. Cuthbert Featherstonehaugh, a famous manager of *Brookong* at Lockhart in the Riverina in the early days, later recalled: 'In 1867 or 1868, when attending what I think was the first show held at Jerilderie, Joe Weir, who used to manage *Yanko Station*, asked me to go down to a bend on the Colombo Creek to look at some rams which were for sale. We both summed them up as not at all what was wanted. They had good frames, but the wool was so coarse that we concluded there must be a long-wool cross in them. Consequently, all the best judges at the Jerilderie show condemned these rams. They were the property of G Peppin and Sons of *Wanganella*.'[6]

Despite the misgivings of these men, it soon became apparent that the Peppin sheep were more profitable to run than other strains, particularly in the dry inland country. Their robust constitutions meant fewer mortalities, their heavy fleeces more than compensated for a lower price per pound, they were more fertile, and could keep producing wool in drought conditions. From 1880 their popularity grew and grew, until by 1930 nearly 70 per cent of all Merino sheep in Australia were directly descended from the stud founded at *Wanganella* in 1861. As John Ryrie Graham said of this pure-breeding strain: 'Such was the surprising durability and vigour of the blood that neither cupidity, neglect, nor total ignorance of the true principles of breeding could destroy the character of the wool.'

The dominance of the Peppin strain shifted the centre of gravity of the stud sheep industry away from the 'inside country'. Offshoot studs of

Wanganella were soon set up in the Riverina and Macquarie Valley regions of New South Wales, where good soils and rich native pastures ensured that sale rams were presented to the market in splendid order. The Peppins had sold out in 1878 to Falkiners, who retained the blood in their *Boonoke* stud, Austins, who eventually retained the *Wanganella* stud itself, and Millears, who named their portion *Wanganella Estate.* Other Riverina-based offshoots were *Barratta*, owned by the pastoral house A.M.L.&F., *Murgha*, owned by the Austins, and *Uardry* and *Burrabogie* of the Mills family. In the Macquarie Valley there rose up two major studs based on *Wanganella* blood – *Haddon Rig* at Warren, later owned by a branch of the Falkiner family, and the Body family's *Bundemar* at Trangie.

The performance and popularity of Peppin sheep were such that these studs developed large businesses sending rams across state and climatic boundaries to all parts of the nation. In 1920 *Boonoke* sold 8259 rams for £68,658; their 150 customers included woolgrowers from the 'inside country' at Rylstone, Bathurst, Singleton, Yass and Gunning, in addition to big consignments to the back country. These included 300 each for *Malvern Hills* and *Portland Downs,* another 300 for *Listowel Downs* and 400 for *Evesham* – all in the Blackall/Longreach area. There were 700 for the Queensland properties of the QN Pastoral Co, and 1437 to the properties of the Scottish Australian Company.

The various stud masters did different things with the *Wanganella* genes – the *Boonoke* sheep became noted in the back blocks of Queensland for their heavy fleeces and hardy constitutions; *Haddon Rig, Uardry* and *Wanganella* sheep for the purity of their wool, and *Bundemars* for their near-perfect conformation. These differences were more in style than substance, more apparent than real. Had the wool of the various strains been measured objectively there would have been little difference, but the visual variations were used as tools in intense competition for ram sales. Breeding and classing methods, though there was no great mystery about them, were kept as closely guarded secrets, and it was not until the 1980s and 1990s that any comparative production information on the performance of the various studs became available.

The great stud masters used every marketing tool at their disposal. Old Franc Falkiner induced his clients to write and tell him how good his sheep were, and pasted their testimonials into a great leather-bound book, which he thrust in front of prospective customers. They all leaned hard on the big wool broking houses, and allocated their wool clips, merchandise orders and livestock sales to whoever bought them the most reciprocal ram business. They employed roving representatives, set up regional ram depots, pressured the influential sheep classers and showered commission and financial inducements round like confetti. They advertised widely in the weekly and monthly journals. In later years television and radio were used, and at *Boonoke* during the 1980s we made extensive use of mailing lists and targeted direct-mail drops.

The stud masters were passionate about their sheep, and intense personal rivalries developed. Neilson Mills of *Uardry* and later *Burrabogie* was fond of saying there were only two people in Australia who knew anything about sheep, and they were himself and Otway Falkiner, and he remained unsure about Otway's judgement.

Until the early 1970s, when studs threw their gates open for the round of Riverina stud field days, no stud master would dream of stepping onto another's property or of looking at their sheep in a pen, and they placed similar restrictions on their employees. In 1956, when I was manager of the Falkiner's *Boonoke North* property, my father was relieving manager at the Australian Estates' *Oolambeyan* stud about an hour and a half's drive away. Though we had not seen one another for two years, he steadfastly refused to visit. It was just not the done thing. We had to meet in the nearby town of Narrandera. For many years jackaroos at *Uardry* were expressly forbidden to mix socially with the *Burrabogie* jackaroos. In April 1961 many of the leading Riverina breeders were somehow induced to attend the centenary of the foundation of the *Wanganella* stud. There is a historic photo of a number of these patriarchs gathered around Otway Falkiner, who was by this time in a wheelchair, all pretending not to notice one another. During the function Neilson Mills, by then also quite elderly, was heard to ask in his high-pitched and penetrating voice, 'Where does a man have a piss around here?' When told he should make use of a nearby myall tree, he exclaimed: 'That's good, I've always wanted to piss all over *Wanganella*.'

Competition went beyond the intense commercial warfare between the various offshoots of *Wanganella*. Pre-dating the Peppins, the Murray and Hughes families in South Australia had been developing their own strain of big, hardy and strong-wool sheep to suit the arid lands. A verse of the time went:

When Mary found her wrinkly yank and nice plain ram a failure,
She promptly pacified the bank with rams from South Australia.

A cluster of parent and daughter studs breeding this unique strain of Merinos became established in the mid-north region of the state. Key players included various branches of the Hawker family's *Bungaree*, the Hawkes family's *Koonoona* and the Collins family's *Collinsville*, which straddled the Peppins and South Australians in type. Though the leading studs of the two strains discretely exchanged genetic material from time to time, in public they treated one another with contempt. When opening the fleece of a strong-wool *Bungaree* sheep, a Peppin breeder might jerk his hands back in mock pain, examine his fingers for imaginary blisters, and exclaim: 'His wool is so dry and crackly that you need to be earthed to touch him.' When during the 1970s it was pointed out to Keith Richardson, a Western Australian breeder of *Collinsville* sheep, that early in the century his stud had been founded on Peppin blood, he responded: 'Yes, there's some Peppin in them, but not enough to do them any harm.'

In 1930 Peppin studs ruled the roost, claiming a 70 per cent share of the national ram market, with South Australian bloodlines taking 20 per cent, but by 1980 the Peppins had slipped to 46 per cent, with the South Australians rising to 42 per cent. This shift came about largely through development of new land and increased sheep numbers in Western Australia. There were 102 local Merino studs supplying the Western Australia market in 1930; by 1970 it had grown to 542, the majority of which were based on South Australian blood. This was not so much a result of South Australian sheep being particularly suited to the new pasture-improved land – the major influence was finance. The dominant pastoral house in the region was Elder Smith and Co., which was based in Adelaide and was the major supplier of working capital to woolgrowers in both their home state and the emerging Western Australia. Many of the large South Australian studs were heavily indebted to Elders during the post–Second World War years, and all parties saw development of the Western Australia ram market in their favour as a convenient way to reduce their exposure. Many of the settlers on the new land were inexperienced, knew little of sheep breeding and marketing, and relied upon their local stock agent for advice, and this situation enabled Elders to sponsor sales of South Australian blood rams, much to the chagrin of the Peppin breeders.

CHAPTER SIXTEEN

WOOLLY-MINDED FADS

Fashion, though Folly's child, and guide of fools,
Rules e'en the wisest, and in learning rules.
GEORGE CRABBE (1754–1832), *THE LIBRARY*, 1808

The battle for market share between strong-, medium- and fine-wool Merino studs waxed and waned according to prevailing fashion in the sheep industry.

Wool growing is undertaken as a commercial enterprise, so the methods used for breeding sheep should be directly related to economic considerations and selection programs directed at improving profitability. Sadly, this has not always been so, and the art of breeding sheep has been subject to whims of fashion every bit as irrational as platform shoes, bare midriffs and flared trousers. The tough pioneers who opened up the inland knew little about sheep but a good deal about making money, so their sheep breeding aims were simple and effective. They wanted hardy sheep that would stay alive and produce heavy fleeces. As the industry moved into the 20th century breeders turned down many strange and devious pathways, for reasons not always related to making money.

Of all the perverse fashions that have over the years taken hold of the Merino industry, none was more destructive than the craze for wrinkled sheep that ran from the late 1890s to the early 1900s. It was begun in 1889 by the pastoralist Sir Samuel McCaughey, who at the time owned the biggest flock in Australia. Always on the lookout for ways to increase the productivity of his million or so sheep, McCaughey concluded that fleece density – more fibres per square inch of skin – was the answer. Seeking the densest rams available, he purchased stock from the state of Vermont in the US, where there had developed a theory that sheep with wrinkled skins were denser than plain-bodied animals, and had more skin area. The animals he imported were excessively wrinkled. So successful had McCaughey been that many Australian woolgrowers took the view that 'if it works for Sir Sam, then it will work for me'. Wrinkles themselves, rather than the increased density and heavier fleeces they were supposed to confer, became the focus for selection. A sort of collective lunacy gripped Australian woolgrowers, and a craze for wrinkled sheep swept the land.

The animals that resulted were grotesque. Pictures of champion rams from the early 1900s reveal unfortunate creatures with weak constitutions, pinched little noses peeping out from faces woolly as a mop, and spindly legs emerging from a great rolling mass of oily folds and wrinkles. Their general appearance was rather like that of a poorly made concertina on the compression stroke, and the fleeces they grew were short and redolent of thick yellow grease and yolk.

Vermont sheep did have their critics at the time. The old Merino men objected to their yellow, greasy wool, to which the snake-oil salesmen of the day replied: 'Don't worry about that – it's as yellow as Californian gold.' Others complained that the Vermonts often carried clumps of hair on top of their many skin folds, for which Horsfall of *Widgiewa*, a strong advocate for wrinkles, had a ready answer: 'Hairy fibres denote masculinity.' The sheep classer Thomas Shaw, a supporter of the old Peppin sheep, responded: 'If hair denotes masculinity, then all I can pray is that my young rams will show signs of virginity.'[1]

If the 1898–1902 drought, the worst Australia has ever seen, had one positive effect it was in finding out the constitutional weakness of wrinkled sheep. They died in droves. The fashion soon ran out of steam, with a return in demand for plain-bodied animals of the old Peppin type. A sad legacy of the Vermont craze was the spread of sheep blowflies, which to this day contribute to around 20 per cent of station running costs in terms of lost production and preventive measures. The smooth-skinned old sheep were unattractive to flies, but the moisture held between the wrinkles and the high grease content of Vermont wool were blowfly heaven, and offered the harbour that allowed the fly population to grow and become a national scourge.

The craze for wrinkles was triggered by the very best of intentions – to increase productivity – but woolgrowers then neglected to effectively measure any gains. Once begun, the fashion, as fashions tend to do, fed on itself and became the stamping ground for carpetbaggers until all good sense was lost. Sadly, this was to be repeated during subsequent sheep breeding fashions, but fortunately none would prove so destructive.

To understand the commercial implications of later fashion changes it is helpful to understand something of the complicated genetic linkages in Merino sheep. These dictate that if you breed for one economic factor, other factors will change concurrently, often in an undesirable way. Thus if you select for frame size, fleeces will tend to become stronger; conversely, if you select for fine wool, sheep tend to become smaller. Selecting for heavier fleeces results in stronger wool, and selecting for finer wool results in lighter fleeces. So a breeder who works towards the sensible objectives of producing a large animal whilst increasing fleece weight and growing finer wool is entering a vigorous fight with nature, and needs to be extremely patient, perhaps more than a lifetime, to achieve his objectives.

A craze for wrinkled sheep swept the land in the early 1900s *(Peppin Heritage Museum, Deniliquin, NSW)*.

For most of the 20th century the market paid insufficient premium for fine wool to compensate for the lower fleece weights of fine-wool sheep, so breeders chased heavy-cutting sheep, and the fineness of the Australian clip slowly declined. The process had begun as early as the middle of the 19th century. South Australian strong-wool breeder John Hughes wrote of a visit to Camden Park and a rather tactless exchange with William Macarthur: 'I asked him why he bred such sheep when mine, in similar climate and country, yielded nearly double the money per fleece. He told me, with an air of pride, that they bred the pure blood. I replied that in our part of the world we bred for the pure money, which he characterised as a Yankee way of looking at it.'[2] In 1927 the Yorkshire wool expert SB Hollings spoke at a meeting of the Bradford Textile Society, and claimed that over 10 years there had been a 20 per cent deterioration in the quality of the Australian clip. In 1949 the Australian Wool Realisation Commission, who had charge of selling down the stockpile of wool accumulated during the Second World War, revealed that since 1940 the proportion of strong wool (23 microns and upwards) in the Australian clip had more than doubled – from 14.3 per cent to 30.8 per cent. Over the same period the amount of medium and fine wool (22 microns and finer) had decreased from 54.6 per cent to 37.6 per cent.

In the middle 1950s, economic work by CSIRO (Australia's Commonwealth Scientific and Industrial Research Organisation) showed that, with the wool market where it was, fleece weight was four times more important in determining returns per head than fibre fineness, and 12 times more important than length, colour or character of the fleece. So it was little wonder that ram buyers sought to select the heaviest cutting rams they could afford, and tended to disregard wool quality, and that the studs responded by putting a high priority on increasing fleece weight. They were responding commercially to market signals. Application of the Wool Reserve Price Scheme from 1970 to 1991 reinforced the trend by offering few premiums for fine wool.

During the 1960s the situation was compounded by a craze for bigger sheep, triggered by growth in the export of live sheep from Western Australia to the Middle East, which required mature wethers of great size and weight. To cater for the export trade, woolgrowers demanded Merino rams of exceptional size and the potential for quick maturity. At the time Western Australia had the fastest growing flock and strongest demand for rams, so studs right around Australia responded by breeding flat out for bigger sheep with longer bodies. It was the heyday of the South Australian bloodlines, and ram buyers at the Peppin blood studs complained: 'Get them bigger, or I will have to go to South Australia next year.' There was a certain lack of good sense in the fashion, which lasted for 20 years, because the sale of surplus sheep represented only a small part of growers' revenue, which was still dominated by wool. Big sheep were not necessarily the heaviest cutting sheep.

A day of reckoning had to come, and when it did it was with a vengeance. When the Reserve Price Scheme collapsed in 1991 the majority of the 4.7 million-bale stockpile consisted of medium to strong wool, leaving a massive overhang in the market for these types. This overhang, and consumer demand for lighter weight fabrics that required fine wool, triggered healthy premiums for finer types. The premium paid for 19 micron over 21 micron rose strongly through the decade to reach 800 cents/kg clean by 1999. Caught in the deepest and most prolonged crisis the industry had ever seen, growers reacted by rushing to fine their flocks down, attempting to capture the premium. They were trying to stay afloat, and the results were spectacular. In 1991 the largest micron category in the Australian clip was 23; by 2000 it was 21. In the early 1990s Australia was producing nearly a million bales of 23-micron wool; by 2000 it was down to 320,000. Production at the strongest end of the Merino category, 25 and 26 microns, was down from 446,000 bales to 120,000. The proportion of the clip in the fine and super-fine categories was still rising in 2001 – in Victoria alone the part of the clip that was 19.5 microns and finer moved from five per cent in 1990 to 23 per cent in 2000.

The efforts of growers to fine up their clips caused huge shifts in market share between Merino studs. Most suppliers of fine-wool rams were relatively small studs located in temperate climates, whereas the traditional suppliers of medium- and strong-wool rams were big operations located in the semi-arid Pastoral Zone. These included the huge Peppin and South Australian blood parent studs that had dominated the ram market since the 1890s. Their power and influence was broken, perhaps forever. The geographic pattern of ram supply was returning to where it had been in the 1860s and 1870s.

The wisdom of attempting to grow fine wool in the Pastoral Zone awaits commercial judgement. The natural environment has changed little since the 19th century, and the same problems that triggered development of Peppin and South Australian sheep could arise again. Even in the late 1990s, when the premium for fine wool was at its peak, comparative analysis showed that the most profitable flocks in the western Riverina were those producing 21 to 22 micron wool. By the middle of 2001 the wool stockpile that had haunted the industry for a decade had gone, and prices for middle micron wool relative to fine wool were rising. The high premiums paid for fine wool had virtually disappeared by 2002 – to indicate the fickleness of sheep fashions, and the folly of trying to chase them, during 2002 strong wool was actually selling at a premium to fine wool. The value of sheep meat was rising too – surplus animals that were struggling to make $10 a head in the 1990s were returning $100 or more, a trend which favoured the large-framed Peppins and South Australians. There were also early signs of a change in consumer fashion, away from the lightweight 'cottony' look towards more robust textures in apparel fabrics, which would create demand for stronger wools. The days of the big, handsome Peppin and South Australian sheep – producing 6 to 8 kilograms of 21–22 micron wool on a 60-kilogram frame, and providing 100 per cent of lambs – may not yet be finished.

CHAPTER SEVENTEEN

THE 'BIG SIX'

All rising to great place is by a winding stair.

FRANCIS BACON (1561–1626)

There developed in the stud sheep industry a number of powerful personalities; charismatic people who dominated their era. Peter Forrest described the type when writing about the early station managers: 'They encouraged the growth of a cult of personality about themselves; they accumulated personal idiosyncrasies so as to nurture legends which would inspire awe and secure their central places in pastoral industry folklore.'[1]

The great stud sheepmen possessed an odd mixture of qualities; they were part sheep gurus, part industry statesmen, and part super salesmen. They held enormous authority regarding anything to do with sheep, animal breeding, wool, and the pastoral industry generally. Few were brave enough to challenge their opinions and judgements on these topics. They encouraged an air of mystery about the arcane practices of stud sheep breeding which, in an age before performance could be measured objectively, added to the mystique that surrounded them. Some carried rough edges, but they tended to mix at the highest levels of power in the commercial and political world of their time, and they wielded great influence.

HS Schollick of *Wanganella*, Neilson Mills of *Uardry* and later *Burrabogie*, and Otway Falkiner of *Boonoke* dominated the era between the First World War and the Second World War. The towering figures of the period following the Second World War were BM Clapham, manager of *Uardry* and later general manager of the F.S.Falkiner empire, GBS Falkiner of *Haddon Rig*, and AL Collins of *Collinsville*.

HS Schollick joins the elite group not because he was a charismatic personality – in fact he was of a retiring nature – but by virtue of being a good shepherd, the sales record of *Wanganella* sheep at the premier Sydney ram sales during the 1920s, and his association with 'Ballymena', the most famous ram of the era. The Austin family, together with the Millears, had purchased the *Wanganella* stud from Peppins in 1878, and in the years before the First World War had enjoyed only indifferent performance in the marketplace. Schollick was of German extraction, and bore traces of a guttural accent throughout his life. Prior to his appointment as manager of *Wanganella* in 1915 he had been a selector in the Nyngan district, and managed properties in Victoria and the *Wandook* run, which adjoined *Wanganella*. He loved his work at *Wanganella*, and in 1916 wrote to Harry Austin: 'I am very happier [sic] than I have done for many years in my new life.'

Schollick was an austere man, well educated and with a taste for art and music. He had a good head for business, but his life really revolved around sheep. Joan Austin Palmer recalled: 'Schollick really loved them; one Sunday morning a visitor strolling near the ram sheds saw him lying in the shade with his head on the side of one of his favourites, both taking a morning nap.'[2] It was this close affinity with his animals that enabled Schollick, through good husbandry and skilful feeding, to bring out their finest qualities when presenting them to the market. There is a famous picture of Schollick taken in 1924 on the banks of the Billabong Creek at Wanganella. He is leaning with his disabled left hand on a canoe tree, a long bamboo pole in the other hand, wearing a tweed suit complete with waistcoat, tie and wing collar. In front of him are lined up 11 superb stud rams, as alike as peas in a pod. The picture became widely known as 'The Australian Eleven'.

Five of these rams were sired by Special Stud ram number 9.1, known as Ballymena. At the 1924 Sydney ram sales one of them topped the market at 3050 guineas, sold to the *Barratta* stud, and the others made 600 guineas, 500 guineas, 250 guineas, and 150 guineas. These were phenomenal prices by the standards of the day. During the two years prior to 1924 a total of 22 rams sired by Ballymena were sold for £17,348 at an extraordinary average of £789 each. This was an amazing record before the days of artificial insemination. Then to cap it off, old 9.1 himself was bought as a five-year-old for 5000 guineas by McCaughey of *Coonong*, where he was named Ballymena and was still siring lambs as an eight-year-old in 1927. A visiting correspondent reported that year: 'His appearance, condition and action are more like a ram three or four years old.'

Wanganella had secured the top sale price at Sydney sales in 1919, 1920, 1921, and the top average in 1922. Their vintage year of 1924 was followed by best average of £656.5 for 18 rams in 1925. Even in the Depression year of 1929 they secured the top price of 2900 guineas. Schollick retired in 1940, and died in Melbourne on June 24, 1944. Forty years after seeing them, old sheepmen like Tom Culley of *Wonga* remembered the 'Wanganellas' of Schollick's era with affection as the very peak of perfection in wool quality, uniformity and prepotency. They were the benchmark by which all others were judged.

Andrew Ainslie Neilson Mills was born in 1881, one of four sons of the Scottish-born Charles Mills, who had bought *Uardry* on the Murrumbidgee River near Hay from Wragge and Hearn in 1876. A stud had been founded there in 1864 with 5000 ewes and a complement of rams purchased from Peppins at *Wanganella*. Neilson Mills took over the stud work under his eldest brother in 1900, and assumed full management in 1908. Mills soon established *Uardry* as a distinct type of the Peppin breed – carrying extremely dense fleeces of very stylish wool protected by plenty of clear grease. Ram sales flourished, and by 1930 he had paid off a £100,000 mortgage hanging over the business, and purchased *Burrabogie* on the opposite side of the Murrumbidgee. By then *Uardry* was selling nearly 4000 rams grossing up to £100,000, second only to the mighty F.S.Falkiner group, and had many daughter studs scattered around Australia.

Mills was a distinguished-looking man, tall and well-built with a large, handsome head. His general appearance, intellect and mannerisms were those of a bishop, but the episcopal illusion was soon shattered when he spoke – his voice was high-pitched and he was overly fond of sarcastic and at times hurtful banter. Over formal morning tea in the stately homestead, with delicate little sandwiches served by a housemaid in black dress and frilly cap, he might ask some ill-at-ease backwoodsman: 'I suppose you wish to buy plain-bodied rams?' If the unfortunate guest answered in the affirmative, Mills would snap back: 'Yes, yes, yes, it has taken me a lifetime to breed the wool onto them, and you will breed it off in a generation.' If his anxious wife pulled him up he would reply: 'That's all right about you dear, you know how you talk about them when they're gone.' I once saw Mills striding his imperious way between the pens at the Sydney ram sales, flanked by courtiers. Drawing up alongside the worst-looking sheep in the shed, he rested his walking cane on the rail, paused for dramatic effect, and announced in haughty tones: 'Ah – and *this* sheep belongs to my greatest critic.' He could quickly take dislike to any newcomer who professed to be a sheep expert, particularly if he was from a research or scientific background. I saw one unfortunate fellow selecting rams for clients at *Burrabogie*, and every time he approached a sheep to inspect the fleece, Mills would announce in theatrical tones: 'Open up, give him light, give him room, give him air – here comes the professor.'

Notwithstanding an aloof and haughty manner, Mills was a talented sheepman, with a great eye for detail and phenomenal memory. The sheep classer Tony Watson later recalled: 'Once we had 50 of the young reserve rams in a race; he went along nominating each one's sire, while we followed up and checked the ram's ear label. He got 43 of them right.'[3] Mills read widely among all the then recognised works on genetics and sheep breeding, and was fond of quoting, 'Be thou diligent to know the state of thy flock and look well to thy herds.'

Perhaps it was looking too closely at herds that got him into a spot of bother when in the late 1930s the family shareholders sued him. Mills was managing director of the family company, Charles Mills (Uardry) Limited. In the early 1930s he had used a private inheritance to buy the property *Wahwoon* – which adjoined the company property *Burrabogie* – on his own account, and founded a *Uardry* blood stud upon it, purchasing 11 rams and 4074 ewes from the company. As the decade wore on it slowly dawned on the shareholders, who took little interest in the business and left everything to Neilson, that the reputation and sales performance of the *Wahwoon* sheep was increasing at a greater rate than *Uardry*. Sinister rumours began to circulate about top sires being spirited over the Murrumbidgee in the dead of night by rowing boat.

There are few fights as vicious as family fights, and during 1937 the shareholders took legal action against Neilson in the Supreme Court of Victoria. They claimed that he had, as managing director, organised the issue of bonus shares as dividends in such a manner as to increase his holdings and voting rights, with a view to eventually taking majority control. The court found against the shareholders, as did the Full Court of the High Court on appeal. But the nervous shareholders persisted, and opened another action in the Civil Court of Victoria in 1938. They sought return to the company of a salary increase Neilson had awarded himself from his agreed entitlement of £2000 a year to £3500 in 1936. They lost on that count, but won another arm of their claim – that Neilson had placed his personal interests as owner of the *Wahwoon* stud in conflict with his duty as managing director of the family company. They sought an injunction restraining him from carrying on the business, and return of any profits to the company. In granting the injunction Justice Duffy found that Mills had nurtured *Wahwoon* under the shade of *Uardry*, had used the reputation of *Uardry* to sell *Wahwoon* rams, and in fact had sold *Wahwoon* sheep to former customers of *Uardry*. Eventually the strife-torn family reached an agreement in June 1938, under which Mills would retire from the company, taking 56 per cent of the sheep and the assets, including *Burrabogie*.

The *Uardry* business was decimated; ram sales plummeted from 4000 in 1930 to 700 in 1940, and it would take all the skills and acumen of BM Clapham, who the family appointed in 1940, to rehabilitate the business. Mills carried on at *Wahwoon* and *Burrabogie* with great success until his mental and physical powers began to decline in the late 1950s. By sheer force of personality he had entrenched such a strong culture and set of breeding principles that it was nearly impossible for anyone to create change so long as he was alive. In the early 1960s his successors were still maintaining the sire lines established by Neilson years before – they took the old wisdom for granted. The sheep failed to move with the times, lost size and quality in relation to competing studs, and the studs were dispersed at public auction in October 1966. It was the final blow for the old autocrat, who passed away in January 1967.

Neilson Mill's great contemporary and rival Otway Falkiner was without doubt the greatest ram salesman of all time. Born in Ararat on December 16, 1874, Otway was the eighth offspring of F.S.Falkiner. He suffered from asthma and bronchitis as a child, and it was the medical advice that unless he was shifted to a dry climate he might die that prompted the family to move to the Riverina and buy *Boonoke* in 1878. He was to outlive all his siblings, and by the time he died aged 87 in 1961 he had become a legend and a giant of the pastoral industry. Otway's first job for the family was managing *Pericoota*, on the Murray River near Echuca. He shifted to *Boonoke* on his father's death in 1910, and became chairman and chief executive of the family company when his eldest brother Bert, later to become owner of *Haddon Rig*, sold his shares in 1915 under rather acrimonious circumstances.

The sickly child grew into a nuggetty, vigorous, energetic man of great physical strength. He was quick-moving, decisive and fearless when faced with a critical problem. He devoted his formidable energies to expanding his firm's ram business, and took marketing as his personal responsibility. Up until the 1950s a Falkiner ram customer, whether buying five rams or a hundred, had to contend with Otway. The *Boonoke* sheep developed a well-justified reputation for constitutional vigour, and Queensland, where ram orders were big, was Otway's favourite stamping ground. For many years, if you lived north or west of Longreach, *Boonoke* sheep were your first choice; they were renowned for their ability to hold condition and keep producing wool on low-protein Mitchell grass pastures. Otway became known as 'The King of Longreach' because of his rumbustious behaviour and legendary sprees at the Longreach Club during his frequent visits.

Otway's frankness and open manner appealed to back-country men of the time, as did his reputation for a certain amount of wildness and excessive behaviour. Suzanne Falkiner said of her family: 'Perhaps the most important complicating factor was the unequal distribution of certain family characteristics: some members of the family were a little wild, others possessed a slight tendency for Puritanism.'[4] There could be little doubt which side Otway's genes came from – he had a sharp and twinkling eye for beautiful women and handsome racehorses. It was widely rumoured he had left a number of 'chance-get lambs' around the country; certainly I knew a few men who wore a remarkable resemblance to the old man. But Otway never allowed pursuit of pleasure to interfere with his drive to breed bigger and heavier-cutting sheep and make another sale. Like others of his era he held an enormous authority that made selling an easy proposition. They lived in an age when authority was seldom challenged and traditional wisdom was widely accepted. If Otway were to tell a customer to buy this or that type of ram, his reputation was such that few dared to disagree with his judgement.

Otway, who liked to be referred to as 'The Boss' and was widely known as 'The Daddy of Them All', took great care in selecting those customers whose stock, by virtue of their country or their management, would act as testimonial for *Boonoke* sheep. He could be blunt in ridding his client list of poor managers. In 1924 a customer from Boort in Victoria wrote complaining about the quality of a hundred rams he had bought the previous year. Otway replied: 'I could see when you selected the rams that you didn't know what you were about, and offered you the services of our sheep classer, which you wilfully spurned. I will not replace any of the rams in question, and so long as I am alive will never again sell you another ram.'[5] Fred Knight told of another incident: 'Trouble came next day. My uncle drove the buyer to the yards, and I went with them to open the gates. They went into the yards, but I stayed in the car. After a time, they returned without speaking, and this silence continued all the way home. It was broken only when we reached the inside of the house. The guest was given a glass of whiskey and curtly told the car would be round in a few minutes to take him into Echuca in time to catch the Melbourne train. When he had left, I asked what had happened. The reply was: "I'll tell you what happened. The bugger offered me three guineas for four-guinea rams."'[6]

The 1920s were some of Otway's vintage selling years, and in 1925 the company sold 10,306 rams for £95,321. Even in 1950, with Otway feeling all his 76 years, the company sold 11,163 rams. No stud has ever matched that record.

His declining years were sad and lonely. His second wife Una had died in 1948, his brothers and two sons predeceased him, his eyesight began to fail in 1946, and he suffered a debilitating stroke in 1957, which put him in a wheelchair. His youngest son, OMcL ('Mac'), had been embroiled in a messy divorce and his skills, as manifest as they were in some areas, were not of the kind required to take over the running of the family company. Otway's last public appearance was at the *Wanganella* centenary in 1961, and the old patriarch and master ram salesman passed away quietly in his sleep at *Boonoke* on October 23, 1961, five days after his distinguished nephew George Falkiner of *Haddon Rig*.

During his working life Otway Falkiner had sold around half-a-million rams, a figure none will ever equal. In a moving tribute, his successor at *Boonoke* BM Clapham said of him: 'The death of Otway Falkiner removes one of Australia's truly great pastoralists and stud breeders. During the Depression years he gave hundreds of rams to graziers who could not afford to pay for them. He told them they could pay him when they could afford to. Everyone paid.' Otway Falkiner saw the Peppin sheep at the time their popularity first began to grow, and he lived to help them dominate the Australian sheep flock. He eventually oversaw reunification of the original Peppin flocks when his company purchased *Wanganella* in 1958. He founded the first truly commercial Poll Merino stud. He spanned the gap between the rule-of-thumb, learn-as-you-go, hit-and-miss methods of the pioneering days and the use of artificial insemination and objective measurement in breeding programs. He was far-sighted enough to push through the appointment of Clapham to take over and control the family's stud flocks. Clapham in turn was a great trainer of men, and people such as myself who had come under his influence were to run the Falkiner company through two further ownerships until 2000. So Otway's influence on the destiny of his flocks spanned 90 years from when he took over *Boonoke* in 1910 to the turn of the century.

Basil Murray Clapham was the product of an extraordinary childhood. His father Murray Clapham was a brilliant man who had migrated from England to South Africa, then to Australia, where he had a meteoric rise to become general manager in Australia for the pastoral house Australian Estates in 1922. In 1925, aged only 40, he died suddenly from septicaemia caused by his barber nicking the top off a pimple on his neck. Basil was only 14 at the time and a boarder at Melbourne Grammar School. His mother had suffered mild brain damage in a car accident two years earlier, and his sister told of what happened next: 'Within 24 days of father's death we sailed for Durban; our home was gone, and Basil was left not only fatherless but without his mother and sister. He would not see us again for five years.'[7] These events kindled a powerful ambition that was to drive, and haunt, BM Clapham for the rest of his life.

Norman Clapperton, who succeeded Murray Clapham as general manager of Australian Estates, continued to pay the young Clapham's school fees, and on leaving school saw him placed as a jackaroo under the great sheepmaster HS Schollick at *Wanganella*. There the inexperienced young man soon showed a talent with sheep when two groups of unbranded rams were accidentally boxed together, and he was able to separate them by detecting small differences in type. In 1931 Clapperton influenced the Australian Estates to employ him at *Raby*, their Warren-based Merino stud, and in 1932 he was transferred to *Terrick Terrick* at Blackall in Queensland to take charge of the fledgling stud there. At the time most rams used in Queensland flocks were imported from southern states, and the Queensland Government was encouraging a local stud industry by issuing preferential stud leases, which enabled companies like

Australian Estates to retain land that otherwise might be resumed for closer settlement.

The young Clapham quickly revealed characteristic adroitness in both commercial and political matters. Early in his appointment drought overtook Queensland, and competing studs had allowed their ram numbers to diminish during a period of low demand. Clapham, on the other hand, took the position that it eventually must rain and accumulated a large inventory by retaining unsold sheep from previous years, and somehow talked Australian Estates into funding the purchase of supplementary feed to keep them in strong condition. Their patience was wearing thin when general drought-breaking rain fell state-wide, and demand for rams boomed as graziers rushed to rebuild depleted flocks. *Terrick Terrick* was the only stud able to offer immediate delivery, so it beat southern studs to the punch, and quickly became the biggest of the local studs. Clapham had created a commercially successful operation from what had been merely a legal stratagem designed to retain land title.

At the time of his appointment Clapham, with the title stud overseer, had a small staff and reported directly to the *Terrick Terrick* general station manager. He wished to have his own command, and as the stud business grew he fomented such rivalry and animosity between the stud and general station staff that they sometimes came to blows over things like the use of equipment. When I was there as a jackaroo 20 years later there were still old wheelbarrows with 'stud use only' painted in red on their sides. Soon tensions were such that the company had no choice but to set the stud up as a separate unit; Clapham was given an autonomous command reporting direct to head office in Melbourne. After seven years *Terrick Terrick* had become a major player in Queensland and Clapham had acquired a formidable reputation for drive, ruthlessness and business acumen, and he began to look for wider fields to conquer.

In 1939 a letter, simply signed 'Breeder', appeared in the influential *The Pastoral Review*, lamenting loss of the 'correct method' of breeding sheep developed by Thomas Shaw. It went on: 'There is only one young man who seems to me to have the knack of selecting the right sheep. This is young Clapham. I have heard that Clapham is doing well in Queensland. A few more with his eye and ability and we might be breeding an even lot of sheep in Australia.' Around the same time Australian Estates in Melbourne received an anonymous letter accusing Jim Broad, manager of their premier stud *Raby*, of everything from incompetence to malfeasance.

HS (Bert) Schollick and 'The Australian Eleven', sons of the famous sire, 'Ballymena' on Wanganella, NSW *(F.S.Falkiner archives)*.

They asked my father, then managing the nearby *Buckiinguy* for the company, to investigate, but – outraged by the proposition – he flatly refused. Many, including my father, believed that Clapham himself was the source of both letters.

Then in 1940 a call came which offered him his opportunity in the major league. The Mills family, their *Uardry* enterprise devastated by the departure of Neilson Mills, called for Clapham. Here was his opportunity to control one of Australia's great parent bloodlines, and he grabbed it firmly in both hands. By 1950 he had restored ram sales to more than 2000 and recovered the stud's reputation in the benchmark Sydney market, topping the sale in 1951 at the then-record price of 3400 guineas.

Personal tragedy struck with the death of his first wife and, anxious for a change in direction, in 1951 Clapham accepted appointment as Victorian and South Australian manager for the pastoral house Australian Mercantile Land and Finance Company. It was extremely rare in those days for a wool broker to appoint someone from the production side to a senior position. But the pull of the sheep yard eventually proved too strong, and when his old friend Otway Falkiner approached him in 1952

Clapham jumped at the opportunity to assume almost complete control of Australia's greatest stud sheep enterprise.

When he took up the appointment as general manager of F.S.Falkiner and Sons Pty Ltd in 1953 Clapham was 44 and at the very peak of his considerable powers. He was a handsome man with smooth, regular features, full but firm lips and deep-set eyes that revealed sadness and shrewdness in equal parts. He had a strong presence, a deep, well-modulated voice and impressive command of language. He dressed snappily, almost to the point of foppishness – his team once got a nasty start when he turned up at the Sydney ram sales in the 1950s wearing spats. His sophisticated and urbane manner was in marked contrast to the craggy-faced, rough-edged, tweed-coated sheepmen of the day. He had taken great care to cultivate almost legendary status regarding his ability with sheep – the most famous example of which involved his pulling a cull ram off a truck destined for the meatworks, which later became the leading sire at *Uardry*.

He quickly put together a team of enthusiastic young men at Falkiners, many of whom were to later carve out their own marks in the stud sheep industry. I was one of them, and we found Clapham to be a passionate workaholic, intolerant, quick to anger, abusive, impatient, difficult to work for, impossible to live with, stimulating, and at times charming. Because of his explosive temperament we nicknamed him 'General Bullmoose' after an irascible cartoon character of the day. He pushed everyone hard, but could sense breaking points and knew when to back off. Like tough army officers and strict schoolteachers who are also competent, Clapham had the respect of his staff, who enjoyed telling anecdotes of his legendary tantrums. He had the knack of getting the most from people through massaging the sensitive spots which motivated them, and the capacity to open doors in people's minds and convince them that what they were doing, however mundane it might have been, had significance within a bigger picture.

He restored the fortunes of the Falkiner company during a difficult period for the industry, and regained the leadership role it had played during the 1920s and 1930s. But his abrasive ways and inability to tolerate those he saw as fools had earned him the enmity of some family shareholders, and they eventually organised a coup that resulted in his severing all connections with the company in 1966. He retired to manage the considerable personal assets he had accumulated, and so at 55 his skills were lost to the industry. He died in 1976.

Basil Clapham towered over most of his contemporaries in terms of intellect and strategic grasp. He anticipated by 30 years the regionalisation of ram markets, and set up production units in Queensland and Western Australia. He foresaw the growing importance of science and technology, moving the company into being one of the first to make widespread use of artificial insemination, forming an alliance with the sheep and wool faculty at the University of New South Wales to develop objectively measured breeding programs. Clapham was one of the most significant figures in Australia's stud sheep industry spanning the three decades from 1940, and possibly the most significant, but for a variety of reasons his contribution was underrated and downplayed. People sensed his ambition, and it made them uncomfortable. They sensed his intellectual capacity and that he was probably smarter than they were, so distrusted him and looked for underlying agendas that often didn't exist. When he walked away from the industry, people with smaller imaginations and shorter sight allowed his vision to wither.

There was probably no man in the history of stud sheep breeding in Australia, other than HS Schollick at *Wanganella* in the 1920s, who sold more high-priced rams during his career than Arthur Leonard (Art) Collins of *Collinsville*, at Burra in South Australia. Collins' father John, from humble beginnings, founded the stud in 1895, and father and son concentrated on blending the large frames peculiar to South Australian sheep with the superior wool quality of the Peppin strain. Early in the stud's development the family made the shrewd move of aligning themselves closely to the South Australian-based wool broker Elder Smith and Co. This paid off handsomely when rapid development and stocking of new land in Western Australia following the Second World War expanded the total market for rams; Elders became the principal wool and livestock agent in that state and sponsored a number of *Collinsville* daughter studs, which provided a solid base for high-priced sales from the parent.

Art Collins, like Schollick, was a perfectionist who stayed close to his sheep, and was the supreme master of his day in offering them to the market, year after year, superbly prepared and in glowing good health and condition. Collins had the good looks and dignified carriage of an ambassador. Tall, with chiselled features, dark eyebrows and a thick hatch of white hair in his later years, he had that extraordinary trait common in many great salesmen – others sought his approval and were anxious to please him. His manner was at times imperious, and he had the knack of extracting from a ram buyer all the money he had to spend. Collins would listen attentively when an agent or sheep classer was introducing a prospective customer, and would then lift his eyebrows and ask, 'And what manner of moneyed man is he?'

Collinsville was a tightly run ship, and its breeding methods and operational practices were closely guarded secrets, but Collins kept open house in the homestead and was an expansive and affable host. The whiskeys he poured for visitors were legendary. In my dealings with Art Collins I never knew him to criticise outright another breeder's sheep, but he would on occasions damn them with faint praise. I once heard him, when asked to comment on an outstanding Peppin-blood sheep, respond, 'Yes, not a bad animal for a *small* sheep.' Nor did I ever see him look at another breeder's sheep in the sale pens, unless it belonged to a *Collinsville* daughter stud.

During the 1960s and 1970s *Collinsville* consistently achieved the top price at the Perth ram sale. One year a local breeder named Ernie Doncon had just paid a record price for a *Collinsville*, and in the bar afterwards was extolling the ram's virtues to an old friend, Tom Scanlan. He went on at great length until, feeling a little guilty about his volubility, he paused and said, 'But of course the sheep is not without a few faults, Tom'. Scanlan, who was something of a philosopher, replied, 'Don't worry about them Ernie; you only buy the good bits, and they throw the bad bits in for nothing.'

When Art Collins died aged 72 in 1969, *Collinsville* was clearly the premier South Australian stud. It had established an outstanding record for selling high-priced rams to other studs, and at the time held the record auction price for a ram of $26,250, achieved in 1962. During the next 20 years the prices for Collinsville rams were to reach undreamed of levels, all based on the high standards of breeding and presentation established by AL Collins. Sadly, a later owner of the stud was to trigger a massive loss of confidence in the auction system through market manipulation and doubtful practices.

George Brereton Sadleir Falkiner, usually referred to as 'GBS', was a throwback in the way he operated to his illustrious grandfather FS Falkiner, who had founded the family's stud sheep business in 1878. His father FBS (Bert) was the eldest son of the founder, and managed *Tuppal* for the family. Unhappy in the family business when FS nominated his youngest son Otway to control the stud sheep, Bert sold his shares to Otway and another brother, Leigh, in 1915. The following year he bought

Haddon Rig, already a well-established Merino stud located in the Macquarie valley near the town of Warren in New South Wales. The price was £135,000. Bert served two stints as a federal member, for Riverina from 1913–14 with the Liberals and then for Hume from 1917–19 with the Nationals. He was too independent of mind to be a comfortable party man, and resigned in 1919 to concentrate on *Haddon Rig* and a number of directorships. He died suddenly in 1929, leaving *Haddon Rig* to GBS, then aged 22.

GBS had taken a degree in chemical engineering at Sydney University, and throughout his life held a wide range of business interests. These included a parachute factory during the Second World War, a clothing company, a refrigeration and air-conditioning business, an import-export business in partnership with wartime fighter ace Clive 'Killer' Caldwell, and a documentary film unit. He was a pioneer owner and pilot of light aircraft, and donated £100,000 to Sydney University to advance the cause of nuclear science. As a young man GBS possessed sulky good looks, heavy-lidded eyes and a deep intensity of manner. In maturity he developed the squat figure and heavy jowls common to many of his family, but also their fierce energy and his father's wide-ranging and restless intellect. Up until his marriage at 43 to the noted horsewoman Pauline Weir, GBS was Australia's most eligible bachelor, and the media of the day developed an almost obsessive interest in his social and commercial activities, which was to last for the rest of his life.

Falkiner took business, and to an extent himself, very seriously. At one Sydney ram sale during the 1950s, the wool firm A.M.L.&F. Co. were auctioning a big line up of *Haddon Rig* sheep, when suddenly the plank above the sale pens collapsed, dumping GBS into the pen with the ram being sold. Auctioneer Alf McGeoch, known as the 'Black Prince' and very quick with his repartee, didn't miss a beat with his sales spiel and quipped: 'Now we are offering two rams, and we know for sure they are both good workers.' *Haddon Rig* refused to offer sheep through A.M.L.&F. Co. for years afterwards.

From the 1920s through to the 1980s *Haddon Rig* was second only to the F.S.Falkiner group in the number of rams sold, and equal to *Collinsville* in the number of daughter studs it influenced. Like his grandfather, GBS was an entrepreneur who looked after strategy and left the operational side of the business to carefully chosen managers. The dour and hard-working AB Ramsay (Sir Alexander Burnett Ramsay, sixth Baronet of Balmain) managed the stud from 1931 up until his death in 1965. An excellent operational manager, Ramsay nevertheless lacked strategic skills, and GBS kept that aspect of the business firmly in his own hands. Another man of Scottish descent, Malcolm McLeod, who had been a protégé of the great sheep classer AL Morrison, classed the sheep and controlled the breeding at *Haddon Rig* from 1936 to 1971 and developed a unique type of animal.

In his later years Falkiner spent most of his time in Sydney, where he maintained an office in the city, and lived in a famous Bellevue Hill mansion. He visited the property, where he was always referred to as 'The Owner', at key decision-making times and, though the irascible Ramsay was very much the captain in respect to property matters and day-to-day management, GBS's long shadow was always there. Curiously, the *Haddon Rig* overseer was generally referred to as 'The Mate' and the manager as 'The Captain'. Like Clapham, GBS was a skilled strategist who foresaw the regionalisation of the ram market, and set up satellite studs in South Australia and Western Australia. He anticipated the explosion of sheep numbers in South and Western Australia following the opening up of new land in the 1950s and 1960s, and was a pioneer developer in the south-east region of South Australia. He too recognised the importance of science and technology, and built a laboratory and employed a veterinary scientist at *Haddon Rig*. He was a widely known identity with a huge circle of acquaintances in the social, commercial and political worlds – the constant stream of visiting celebrities through *Haddon Rig* helped maintain a high public profile for the sheep and wool industries.

GBS contracted cancer in 1961, and worked hard and bravely in both private and public arenas right up to his death in October 1961, aged only 54. Yet again a powerful figure was taken from the industry as it approached a time when it was to be in great need of intellectual leadership.

There have been dozens of great stud sheep men, and many would argue that no list is complete without Tom Culley of *Wonga* or Sir Walter Merriman of *Merryville*. But few have exerted the sheer volume of influence as did the big six of HS Schollick, Neilson Mills, Otway Falkiner, BM Clapham, Art Collins and GBS Falkiner. The period since 1990 has seen the market for rams become fragmented and oversupplied, regionalised and localised. Hard information is widely available on the commercial performance of studs and bloodlines, and ram buyers are giving themselves a wider choice. So it is unlikely that a few people will ever again wield the same degree of influence as did the Big Six.

CHAPTER EIGHTEEN

HEYDAY AT THE FAIR

There's no business like show business.

IRVING BERLIN (1888–1989)

Those who attended the Sydney Stud Sheep Fair in the 1950s and early 1960s, when it was held at the old Moore Park Showgrounds, saw an industry at its peak – full of confidence, proud of its achievements, feeling secure about its place in Australian society, and looking forward optimistically to what it believed would be an even brighter future.

The fair was held in early June, a quiet time on most farms and stations, so the week-long event became a mecca for great crowds of woolgrowers from all over Australia. They made a pilgrimage to pick up the latest trends, listen to the chatter and check out the state of the art. Of course a bit of a booze-up, a stay at a nice hotel and a taste of Sydney's fleshpots, all as deductible expenses, had their attractions too.

Accommodation was at a premium, and most people made bookings through their wool brokers. The Australia or Wentworth hotels for the affluent, and the Metropole, Carlton, Ushers, Adams or Royal Exchange for the more frugal. The daring may have chosen the Hampton Court or Mayfair at Kings Cross or the brassy new Chevron in Macleay Street (known as 'The Gin Palace'), and there was the elegant Belvedere in Bayswater Road for families who wanted a quiet time. The wool brokers ran a shuttle service to the hotels, picking up their interstate clients from Mascot Airport.

With few restaurants, only a thin smattering of nightclubs, and archaic liquor laws that closed all public bars at 6 pm, a good deal of social activity surrounded the lounges and dining rooms of the great hotels. There was also a busy round of social events on most nights; sheep breeders society meetings and dinners, parties organised by the studs for their clients, the popular New South Wales Sheep Breeders Association cocktail party and fashion parade at David Jones on the Friday night, and Elder's cocktail party on Sunday night. The Sheep Show Ball, a black-tie affair, was held on the *Showboat* as it cruised the harbour, and was the start of many love affairs. For a few years the New South Wales Sheep Breeders Association made the mistake of holding its annual general meeting at 8 pm one night during the fair, a time that ensured delegates were well tanked-up and ready for a scrap. Meetings tended to be fiery and debate protracted, though some distinguished members slept the meeting through and snored steadily.

From the late 19th century until just after the Second World War, show and sale sheep had minimal preparation other than some supplementary feeding and shedding out of wet weather. In the early years of the 20th century, selling brokers often paid the feed bill for rams offered in their catalogue. But by the 1960s preparation and presentation had become an art form that, to the uninitiated, verged on trickery. It began back on the studs in June the previous year, when rams intended for the shows and auctions were shorn with hand shears, which left a soft blocky tip and about 15 mm of wool on the sheep. A creative blade shearer could 'shape' the animal – disguising wrinkles by shearing the tops of the ridges short and leaving extra length in the valleys. Hollow backs and other physical faults could be dealt with in a like manner. Superior length of wool could be a great advantage on the show and sale floors, so the practice of cheating by shearing earlier than the official date of the first of June became rife, to such a scandalous extent that the Stud Breeders' Associations eventually had to set up travelling inspection committees to check the fairness of shearing.

Rams were shedded out of the weather for nine months before the event and hand fed. Feed mixes were variations around standard ingredients of chaff and oats, with more powerful additives like cottonseed meal, grain that had been put through a heat treatment that boosted its protein content and turned it into a toffee-like substance, lupins, and even apples. Each stud master had his secret mix, and Jim Broad, manager at *Raby* in the 1930s and 1940s, sent everyone away from the shed while he mixed his witch's brew on a sheet of tin. For around three months prior to a show or sale rams received very close attention. A heavy fleece up to

130 mm long tends to obey gravity and fall downwards rather than stand out at right angles to the skin surface, so many hours were spent pulling staples upright with the hands. Wool on the thighs, arms and fronts was teased out with the fingers to make it appear bulkier. The tip of the fleece was broken up to develop a soft and attractive feel – sometimes this was done using a tennis racquet. Sheep with a light tip or dry handle could be smeared with wool dressing, a mixture of lanoline and lamp black, which was worked through the fleece. After a good working up, finely honed hand shears were used to remove fluffy ends and create a softly textured tip that invited touch. Shears would also go into play to disguise things like badly set horns or close hocks. This process might take a full day for a top ram, and be repeated many times in the lead-up to Sydney. Wool tended to clag up and the fibres became entangled when sheep were exposed to the moist, humid air of the city. So it was common to see stud men working on sheep late into the night, with a bar radiator each side to provide heat, while they restored the quality of the fleeces. During the 1960s, when frame-size and bulk of fleece were telling factors in winning show ribbons and achieving high prices, it is almost certain that some studs pumped sheep up with injections of growth hormones. It was said of one stud that consistently exhibited huge animals that 'they are so full of chemicals that they glow in the dark'.

For judging, attendants held sheep side by side in a long line, sometimes 30 or 40 sheep to a class. The process of judging was hardly a riveting spectacle to anyone other than participants, as judges moved up and down the line, opening fleeces, turning sheep over to inspect belly wool, deliberating in conference. It could take hours. Stud men were passionately committed to their sheep, and any judge perceived to be inexperienced or lacking in confidence came under heavy pressure. An exhibitor might sidle up as sheep were being led out, and murmur behind his hand, 'This ram has never been beat at four shows this year, and the last two outings were judged by Sir Walter Merriman.' Others might edge close to the arena to peer over the judge's shoulder as he opened a competitor's sheep, making audible comments as to its quality. If a watching exhibitor detected a crooked foot on a competing sheep, and suspected the judge had not, he might stand so close that the judge had to walk around him, staring fixedly at the offending joint, or pointing and making comments to his neighbour in the judge's hearing. Attendants holding the rams also entered the fun and games. By discretely manipulating an animal's head, by small pushes and pulls, they controlled its posture to display its virtues to best advantage. Spectators could be observed grimacing and pulling at their ears as if demented or drunk – these were prearranged signals to holders about how their ram was standing. When the judge approached, a skilled holder might quietly turn the ram's head towards him and force his knee into its front to compact the fleece and make it appear denser. During the years immediately following the Second World War studs were not allowed to hold their own sheep, and the duty was allocated to students from Hawkesbury Agricultural College. Identifying horn brands were masked with tape. This was supposed to stop judges recognising sheep of a favoured bloodline by knowing the holder or the brand.

Preparation and showmanship had always had its critics. In 1924 AH Wright of Delegate New South Wales wrote angrily: 'You are very mild in your remarks terming it irregularities; it is rather more than that. It is now carried out to such an extent that one is quite safe in terming it fraud and deception. Nowadays to be a successful exhibitor you must also be a successful artificer, or you will have no chance in life.'[1] The noted

Prize rams were cosseted, shown and sold at annual sheep shows *(F.S.Falkiner archives).*

stud man EH Kater, writing for the benefit of his successors in the early years of the 20th century, advised: 'To those who come after me I leave my advice. To the man who shall try to change my ways of breeding, for the sake of adulation at a show, I leave my hearty curse. If I again give way to that man I hope my family will consult a doctor as to my sanity.'[2]

With show business completed on the Friday night, stud men got down to the sharp end of the game, and all sale sheep were in their allotted pens by Friday evening. Saturday was left for recreation, and after feeding and tending their sheep, the pavilions emptied out as everyone dispersed to the races, football, pubs or pursuit of new romances. Sunday was inspection day, and the vast sale sheds were a bustle of activity. Buyers were in the pens from early in the morning checking which sheep suited them. The sheep classers were prominent, selecting for their clients: the aloof and patrician Frank Pennefather, dignified and immaculate in a tweed suit; Harold and Tony Watson, keeping to themselves and diligently marking up buying limits in their catalogues (someone once asked Harold if they could hold his catalogue for him – the look of shock and disbelief that came across his face was something to see); JB Dowling resplendent in a garish tweed sports coat he had made expressly so that he could be seen in the crowd (someone once asked him where he got the material so they could have a dressing gown made from it). There were crowds of busy people wherever you looked – someone once said that you spent half your time looking for people, and the other half not finding them.

Stud masters and their staff patrolled their sale pens, pulling sheep out for inspection, extolling their virtues, cobbling deals together, offering incentive commissions, noting with gimlet eyes who was looking at which sheep, sorting out who were genuine buyers and who were browsers. They kept a sharp eye on their competitors' pens too, and were quick to intercept any regular client they saw inspecting opposition sheep. Shrewd buyers used diversionary tactics, paying undue attention to sheep they were not really interested in and ignoring their real choices. The brokers' stud stock representatives, urgers and commission salesmen bobbed up everywhere, drumming up business for their catalogue or stud, or arranging finance packages. Little conferences formed in the corners of the sheds as groups worked out buying or selling tactics. Salesmen skipped round the pens – 'There is an interesting sheep over here I would like you to have a look at – it has qualities you need' – as they led a potential buyer to the studs who offered them commission or paid their retainer.

On Sunday night stud masters and their selling teams gathered in hotel rooms to review the day, discuss what competition they might expect on each ram, and set reserve prices for the following day. They hoped to work out the bidding limit of each potential client and find ways to extract full value. If a stud felt they had strong interest in a particular ram they might instruct the auctioneer to open the bidding, punt the market along with dummy bids until the reserve was reached, then it was a quick knock and sudden death.

From early Monday morning the atmosphere in the sale sheds, the Commemorative Pavilion and Manufacturers Hall crackled with excitement. Traditionally everyone dressed up for sale day, wearing their best suits. Last-minute inspections were made; vendors gave final instructions to their brokers. At 9.30 am a bell rang and up went the cry of 'Sale-Oh! Sale-Oh!'. From that moment attention turned to the leading auctioneers of the day: the elegantly dressed Alf McGeoch for A.M.L.&F. Co.; 'Gentleman' Bertie Steege and the stentorian-voiced Tim Jones for Goldsbrough Mort; the ebullient Dick Gaden for Dalgety; and breezy John Windsor of rapier wit for Elders. In the 1950s sales commenced on Monday morning and went through until Wednesday evening. Agents sold in rotation, the one who sold first moving to last the following year, then moving back up through the field. To be first sale or last sale was a disadvantage; first because the early market was uncertain until it established a tone, last because demand might dry up. A time limit of just over a minute a pen was imposed, and any agent who failed to complete their catalogue in the allotted time could be 'rung off'.

Prices at Sydney ram sales fluctuated according to the wool market, industry confidence, and the quality of the season. The event began to lose popularity during the 1960s as woolgrowers' terms of trade declined, and in a controversial effort to reverse the slide it was switched from June to a summer fixture. The decline continued: 'The only things you can sell in a Sydney summer are beer and bikinis,' growled BM Clapham. The sale was later shifted to Dubbo in central-western New South Wales, which did little to improve things. In 1977 a total of 380 rams were sold in Dubbo; by 1998 sales had dwindled to 117.

There were a number of reasons for the decline. The excessive amount of preparation put into showing off top rams resulted in their bearing little resemblance to what average woolgrowers saw grazing in their home paddocks. So the event lost relevance, and when things got tough during the 1960s and 1970s, and growers became busier as they ceased to employ labour, they had neither the time nor inclination to spend days away from the farm. Crowds dwindled to consist mainly of the stud people themselves, their hangers-on and agents. So studs found that the audience they were left with consisted mostly of their competitors, and the very purpose of the events – which was to write new business – had been destroyed. Show ribbons also ceased to be a means of competing in the market because commercial woolgrowers lost interest, and in any event they were gaining increased access to performance information of a more commercially relevant nature.

The centralised multi-vendor auction sales also fell into disrepute because during the 1980s the market became corrupted. Sale-topping auction prices generated a good deal of publicity, and studs began to see them as testimonial that they could use to enlist new customers. Sale results were manipulated – a ram might be pre-sold at a certain price, then put to auction and the bidding punted up to a higher and fictitious figure, with the vendor simply rebating the difference to the buyer. Prices paid for some rams were artificially inflated by packaging deals up with all sorts of incentives – generous terms, truckloads of stud ewes, access to semen and other rams, tax-minimisation schemes involving embryo transfers, and so on. The whole package would be built into the price of one ram transacted at auction at an agreed price. Sometimes potential bidders were given guarantees of subsequent semen sales if they bought a certain sheep.

It all came to a head during the wool mini-boom of the late 1980s just before the whole wool market crashed. In October 1988 Queensland property developer Mike Gore, doyen of the 'white shoe brigade' who had recently purchased *Boonaldoon* near Moree, turned up at the Dubbo National Ram Sale amidst a flurry of publicity. The conservative stud men, in their quiet way, summed up the brash newcomer and proceeded to relieve him of a good deal of money. He set a world record by buying a ram from the South Australian *Collinsville* stud for $330,000 and another from *Haddon Rig* for $190,000 (in an interesting reversal of fortunes, several years later the two rams, by then in possession of an AI centre, were offered to the F.S.Falkiner group for $1500). Nothing like it had been seen since the golden days of HS Schollick at *Wanganella* in the 1920s – and there was more to come. At the Adelaide ram sale in September 1989, despite further softening of the wool market, another

Collinsville ram sold for $450,000 to Richard Nitschke, son of the noted state cricketer 'Slinger' Nitschke, of the *Willgoleche* stud. In what newspapers described as a 'football grand final atmosphere', bidding opened at $100,000 and went up in $20,000 steps, accompanied by spontaneous applause. It was interesting that the applause was mixed with hoots of derisive laughter from some sections of the crowd. In March 1989 *Collinsville's* owner, the entrepreneurial West Australian Neil Garnett, whose enterprise was to collapse in spectacular fashion within a few years, confidently predicted that a Merino ram would shortly make $1 million. He based his prediction on new technology that enabled a ram to produce 20,000 straws of semen a year, which would generate $1 million in revenue from sales at $50 a straw.

These events marked the final gasp of the great auction sales. As Paddy Handbury, a later owner of *Collinsville*, said when announcing the stud's withdrawal from Adelaide sale: 'I got a bit sick of people telling me we had to find a buyer for the rams, and we would if we gave them a truckload of ewes.' It was the close of a colourful, exciting, and in many ways romantic era. Raleigh Sellars, a dapper and quick-witted sheepman who represented the Riverina stud *Coonong* during the 1960s, recalled the experience of its foot soldiers: 'Who would want to relive the years of breeding, months of preparation, the days and miles of travel and the final hours spent on detail? All that goes into getting those magnificent animals to the point of final judgement by both the judge and end user. The one that should have been there but didn't make it. The leaking tarpaulin that ruined a champion. The broken tooth the day before, the undrinkable water that hollowed them out, the butterflies and the torment as the chap who was going to buy them fails to show, or spends the morning looking elsewhere.'[3]

By the opening of the 21st century Australia's stud and commercial sheepmen, with absolute good sense, had arrived at a position more akin to the one that existed in the late 19th century, when practices were more aligned to commercial needs of the day. Those who wished to obtain the top genetics were able to purchase semen from proven sires, accompanied by detailed performance information – a much less risky proposition than buying a high-priced ram at auction that might prove infertile or turn out a dud breeder. Most of the auction activity had switched to annual sales on the stud properties, where rams were presented in more or less natural conditions, were a year younger, and were supported by performance information of various kinds. The studs turned over their rams quicker and avoided having expensive sheds and feeding systems that could cost $2000 a head to prepare a ram for sale.

The great capital city shows and sales served a useful purpose in their time. They provided a central meeting place and forum for woolgrowers, and gave the industry an interface with the wider business community of the cities. Their significant contribution in these respects has not been replaced.

CHAPTER NINETEEN

NEW MEASURES

With what measure ye mete, it shall be measured to you.

GOSPEL OF ST MARK

During the 1960s and 1970s a reluctant industry was forced to change its ways by an increased ability to specify the wool fibre and the fleece. So fundamental were the changes brought about that traditional systems and methods that had been around since the days of Macarthur were challenged and eventually swept away. Ancient craft skills – that were almost mystical in nature and deeply satisfying to use, and were often passed from father to son – were found to be inadequate to cope with the needs of a modern industry. The new knowledge made many people feel insecure, and traditionalists manned the barricades to resist advances in technology. The wool industry was to be held back by its rich traditions and history – not for the first or last time.

To understand the strategic importance of these changes it is helpful to understand how wool was being specified for sale, to look at the means by which it was divided into various categories for the purpose of valuation and how the methods of specification were changing. To know those aspects it is helpful to understand how greasy wool that is shorn from a sheep is transformed, through a series of steps, into fabric.

When a new fleece, warm and gleaming ivory white, is dropped onto the shearing board it contains on average 40 per cent of impurities. These are in the form of grease, wax, dust and vegetable matter picked up by the sheep in the 12 months it took to grow the fleece. So, as a first step, all wool is washed in hot water and detergent, a process that removes most of the impurities except vegetable matter. This is called 'scouring'. So the first essential step in categorising or valuing greasy wool is to estimate, by subjective or objective means, the amount of clean wool that will be left after scouring. This 'clean scoured yield' is expressed as a percentage; most Merino fleeces, for instance, will yield 65 to 70 per cent of clean wool.

Scouring wool is processed on one of two systems: the Woollen System that uses short wool to produce a fluffy thread (yarn) for making lovely soft jumpers and other knitwear; or the Worsted System that uses longer wool to produce a smooth yarn for use in high-quality cloth for superb suits that will drape beautifully, withstand soiling, and last for years.

No-one has described these two industrial processes better than CEW Bean in *On the Wool Track*. Though he was writing in 1910, and the technology has since improved, in principle it remains the same. He described the Woollen System thus: 'Wool is not really the smooth, sleek hair that it seems to be. It differs from hair in being sheaved with strong, rough-edged scales. Under a microscope wool looks like the rough trunk of a date palm. When these horny threads are laid across one another, they interlock. Men have invented a machine which takes the wool exactly as it comes from the scour, and then proceeds first to separate it and break up the order in which the fibres have arranged themselves on the sheep's back, and afterwards to relay themselves across one another at every possible angle. It continues to separate them again and again until they are so matted that they come off in the form of a thin, gauzy film – utterly without strength, but still hanging together.' The way in which the machine lays and relays the fibres is by catching them on rollers clothed with millions of small metal teeth, from which other rollers with gradually diminishing teeth pick them off and shuffle them, and pass them on to still smaller rollers. The gauzy strip is then rubbed just like a man rubs tobacco in his palms, so the strips come out rolled, but without any twist. This product is called 'sliver': 'It takes only one act to turn these soft, feeble Slivers into a strong woollen thread. The loose conglomeration of fibres has only to be twisted in order to bind it very firmly, and that is what "Spinning" is. The thin, strong woollen thread which results is the product of a twist given to a sort of barbed-wire entanglement of fibres, a confusion worse confounded. The ends of the fibres stick in all directions on the surface of the thread; these ends make a soft, fluffy covering to the thread, like the down on a peach.'

The Worsted System of processing is entirely different, and Bean describes it thus: 'It takes very little account of the hooked scales. Instead of mixing up the fibres at every possible angle in order to entangle them, it makes a thread by laying the fibres neatly alongside of one another, and then drawing them out and twisting them as string is twisted.' The long wool must first be cleared of any short fibres that may be in it. This is done by combs, which also begin to straighten out the fibres. This process is called 'combing', and the product – a continuous strap of wool with the fibres parallel to one another – is called 'top'. The short fibres that are combed out are called 'noils'. 'The thread is made from combed wool by drawing it out and spinning it. Naturally this thread, in which all the fibres had been carefully laid lengthwise, was very different from the fluffy thread made from the other process. Hardly any of the fibre ends protruded. It was a hard-faced shiny thread.'

Fibre thickness is important in determining the value of greasy wool, because thin fibres produce a light yarn, with many threads in its cross-section that can be knitted or woven into soft, high-quality fabrics that

Spectacular rams bred on *Wanganella*, NSW *(F.S.Falkiner archives)*.

command a high price. Consequently thin (fine) wool is more valuable than thick (strong) wool. Up until the 1960s there was no commercially feasible way to objectively measure the diameter of wool fibres, even though the technology for doing so was around. The method used by wool buyers, valuers and classers was based on craft skills developed in early times around Bradford in England, when that region led the world in wool processing. These skills were used to estimate fibre diameter through observing the frequency of the tiny crimps or serrations that occur along the length of a staple of wool. The more frequent and closer together these crimps, the finer the wool was supposed to be. Softness to the touch was also used as an aid in estimating fineness. Under this method fineness was expressed in numbers, known as 'counts', which were supposed to be the number of hanks of yarn, each 560 yards long, that could be spun from one pound weight of combed wool. Thus one pound of 60's wool would produce, it was supposed, 33,600 (60 x 560) yards of yarn. The higher the count the finer the fibre, so fine wool was either 80's or 70's, medium wool 64's, strong wool 60's or 58's.

Craft skills were passed on through practical demonstration and were taught in schools and colleges. They were used for making critical commercial decisions by people right along the production chain – sheep buyers, stud breeders, sheep classers, wool buyers, traders, merchants and mill owners. Many were far removed from processing activity with no means of auditing or calibrating performance, so variability was great. During the 1960s technology became available on a commercial basis that was able to objectively measure clean scoured yield and fibre diameter expressed in microns – one-thousandth part of a millimetre. The inaccuracy of the old subjective assessments was soon revealed and it became evident that crimp frequency had only a poor relationship to fibre diameter. Speaking at a convention of sheep classers in Dubbo in 1969, Professor Ken Whitely of the University of New South Wales reported on trials that showed that samples of wool expertly appraised at 68 per cent yield in fact had a range of 55 per cent to 75 per cent when measured objectively. With fibre diameter, samples of wool all appraised visually at 58's count actually had a five-micron difference when measured – representing a range from 70's (fine) to 56's (strong).

The ability to measure accurately the two most important factors that determine the value of wool was to have profound implications for the whole industry. It had the potential to streamline marketing systems and reduce labour costs — because wool no longer needed to be present at the point of sale, and could be sold sight unseen. It changed the methods used to select sheep for breeding purposes, improved processing performance, and reduced the number of commercial disputes. It was to be an important weapon in wool's war against synthetics. Yet, as often is the case when revolutionary change is involved, wool people found difficulty in coping with the consequences. They felt their rich culture was being invaded and there was strong resistance to adoption of the new methods. Mills and exporters feared their commercial interests were being threatened, and wool buyers and sheep classers that their sectors would become redundant. Growers, who had the most to gain, were apprehensive and failed to understand the benefits that were to be gained by accurate measurement. Put simply, because of the imprecise nature of the old subjective methods and their high margin of error, wool buyers were inclined to make very conservative judgements about specifications and value, and ultimately this was at the cost of the grower. For example, the accepted rule of thumb was that Riverina fleece wools would yield round 65 per cent – objective methods showed that in fact they mostly yielded 70 per cent. The buyers had been ripping off the growers for years.

In addition to vested commercial interests, there were powerful human factors at play. Assessing wool in the traditional way involved observing it closely, compressing it in the hands, feeling it with the fingertips, sniffing it, holding it up to the light, snapping it between the fingers to test its tensile strength. Experienced wool men could read the provenance and history of every fleece. Colour of the dust indicated what type of country it was grown on, the type of vegetable matter indicated the seasonal conditions under which it was grown, minute variations in crimp told of nutritional variations during the year, purity of flock genetics was read through style and uniformity. Some buyers could even work out the bloodline of rams being used in the flock. Handling wool and making these value judgements was a pleasurable experience in both physical and intellectual terms, and people became passionate about the fibre. So it seemed just too hard to contemplate that all these craft skills learned through years of trial and error, all this tradition, all that passion, could be replaced by precision measurements taken by men in dustcoats in the clinical atmosphere of a laboratory. The clash of cultures was to go on for years.

CHAPTER TWENTY

SCRUTINY OF SCIENCE

He that will not apply new remedies must expect new evils;
for time is the greatest innovator.

FRANCIS BACON (1561–1626)

During the years immediately following the Second World War there lived in the Condobolin district of New South Wales a certain stud breeder who had a standing order with Neilson Mills of *Burrabogie*. Each year Mills would crate up and forward by rail a stud ram at 500 guineas – a considerable sum in those days. There were no specifications other than age and price. The breeder had absolute confidence in Mills' ability to provide value by selecting a suitable sheep on his behalf, and that the sire would increase the productivity of his sheep. The breeder's son, recently back from the war and with ideas of his own, noticed that as each year passed the rams seemed to get smaller, more wrinkled, and cut lighter fleeces. Disturbed, he discussed with his father the notion of trying another stud, or going down to select a sheep themselves and negotiating the price. His father was outraged, and the mere suggestion nearly caused a major family rift: 'You ignorant and ungrateful young fellow! How dare you question the judgement of Mr Mills, who is Australia's leading stud man!'

Such was the reputation and authority of the men who controlled the great parent studs that generations of woolgrowers up until the 1960s treated them with almost reverential awe. But that situation was to change very quickly. In the 1950s and 1960s a group of researchers came up with the not unreasonable proposition – as we saw in the previous chapter – that productive characteristics that breeders selected using subjective craft skills could all be measured objectively, thus greatly increasing accuracy. Fleeces and sheep could be weighed, fibre fineness could be measured. By the 1970s most of the specifications that determine the price of greasy wool – clean scoured yield, vegetable matter and fibre thickness – could be measured with reasonable accuracy, and wool was being traded and sold on those specifications. So it made good sense that sheep should be selected and bred using the same specifications.

Traditional stud breeders and sheep classers contended there were deeper mysteries involved in selecting the best sheep, mysteries that could only be solved using craft skills, and that measurement alone was inadequate to discover the truly influential sires. Some saw their livelihoods as being threatened through loss of the mystique surrounding their craft or profession.

Stud breeding had always been a game of chance, and the eminent geneticist Professor Bede Morris described the gambles breeders made every year: 'When he mates these carefully selected animals, he is faced with the inevitability that the characters so carefully chosen will be re-sorted in an unpredictable way amongst the progeny. The breeder thus becomes the dealer in a card game in which the pack of cards he plays with represents all the genes, good and bad, within his flock. With each mating, the pack is shuffled and dealt into new hands, always in the hope that eventually some hands will come to have all the aces and the royals, while the rags will be discarded. But the aces and royals can only be identified after the progeny are born and grow to adulthood.'[1]

The ability to accurately measure the performance of an animal could only serve to diminish the risk of a poor hand and shorten the odds for getting a good hand. So researchers mounted a campaign for increased use of measurement by the great Merino studs. They spoke in a calm and rational way at meeting after meeting, convention after convention, with masses of supporting information. But reason was insufficient to overcome the stud breeders' passionate defence of the traditional methods they held dear. One of the strongest proponents for change was Dr RB Dun of the New South Wales Department of Agriculture. Dun, in a well-intentioned effort to stimulate industry debate on the issues, was deliberately provocative and controversial. He wrote in 1963: 'There was a vested interest in stud breeding which was to prove completely impervious to scientific thought. The studs have no economic problem. Ram buyers have been well-educated to ask for the product that is for sale. The stud master doesn't want to know that anyone that can read a manuscript and read a set of fleece-weighing scales can successfully breed Merino sheep. Staff, particularly the sheep classers, who have a strong position in the industry, can see a basic threat to their livelihood. It is only natural that they form a strong pressure group against research findings which negate the value of their traditional skills.'[2]

An alternative view came from a respected English scientist, Allen Fraser: 'The facts of today become the errors of tomorrow. Therefore, to place the direction of animal breeding in the hands of scientists seems to me to be to take an unjustifiable risk. The risk is the greater because of the growing dogmatism of science, certainly of the agricultural sciences, to assert unquestioned authority in the ordering of things biological as the medieval church at one time claimed in what concerned the soul. It is much safer to regard science as being one method of seeking knowledge rather than as knowledge itself. For one thing there is nowadays quite as much vested interest in science as in animal breeding.'[3]

The truculent reaction of most stud masters and sheep classers to the new knowledge was much less moderate. Dun had left himself open to criticism by advocating selection programs heavily biased towards fleece weight, without paying sufficient attention to fleece quality, but he really set the packs baying after his blood when he began to act as an advocate for the South Australian bloodlines. Traditional breeders treated much of the work with derision, making coarse jokes about sheep yards being too dusty for computers, people tripping over their slide rules, and so on. An unfortunate consequence of Dun's provocation was that the two sides bunkered down, and a dangerous no-man's-land was created between them. This gap persisted right through to the 1990s, despite myriad conferences organised to bridge it. Adoption rates for the new methods at stud level were snail-like until the late 1990s. Even as late as 2001, when industry research bodies were being restructured, there was political lobbying, led by some traditional stud breeders, to reduce funding for further research into breeding systems and genetics.

During the 1990s there was increasing demand from ram buyers for specifications and performance information for the stock they bought. More information on the performance of various bloodlines and studs was being generated. This exposed the Merino studs to greater scrutiny, and changed the basis of inter-stud competition. As the volume of performance information grew, ram customers were able to base buying decisions on hard facts, and no longer had to rely upon show results, auction performance, the reputation of the various stud masters, or blandishments of ram salesmen. They were less likely to maintain brand loyalty and shopped around for rams.

Triggered by the collapse of the wool Reserve Price Scheme in 1991 and a decade of disastrous wool prices, the Australian sheep flock had fallen 48 per cent from more than 173 million in 1990 to around 90 million in 2003. More ewes were being joined to meat-breed sires as growers sought access to strong sheep meat markets in an effort to stay afloat. So the market for Merino rams was shrinking, and during the 1990s some of the larger studs lost 40 per cent of their sales volume. At the same time the supply of Merino rams had increased. In 1980 there were 1536 Merino studs in Australia, joining 812,500 ewes; by 1995 there were 1825 joining 1,356,777 ewes – an increase of 67 per cent in joinings. The market for rams had become grossly oversupplied, and prices were being squeezed. The ram market was also becoming regionalised as breeders choose to buy acclimatised stock locally, so the great parent Merino studs – which had attained icon status and were rich symbols of Australia's great era of wool – were coming under pressure and losing their earlier dominance. Producing rams had become unprofitable; costs per sheep in operating a stud were twice those of running commercial flock sheep – mainly through additional labour costs for artificial insemination programs, identification of animals, recording measurements, as well as things like advertising and promotion. The increased revenue generated by selling a ram at, say, $400 compared to a wether at $30 was eroded by high wastage rates – it took about 10 breeding ewes to produce one ram for sale.

The once powerful stud sheep breeders' associations, whose committees during the first half of the 20th century had included some of the most influential men in the nation, seemed unable to cope with the complexity of the issues confronting them, and sank into dealing with trivia. Instead of tackling issues such as the mechanics and ethics of specifying their product, or the trading of genetic material in new forms, or the likely impact of genetic engineering, they took refuge in the minutiae of running shows and auction sales, which were themselves becoming irrelevant to most sheep men. Parkinson's Law took over, and a typical meeting might deal at length with provision of morning teas for judges, quality of the PA system, and the placement of hooks on locker room doorways.

As the stud Merino sheep industry moved into the 21st century an old order and its traditions were in decline. The old industry was full of mystery and magic, peopled by rich and colourful characters, by charlatans and statesmen, and enlivened by fierce competition and dispute. The one that replaces it will be more precise, more specified, more controlled, more predictable, more attuned to the commercial needs of its customers. The generation that will be running sheep in the future will be those who grew up in the computer age and are accustomed to analysing information. They will base their ram buying decisions on performance data, which will be evident. In 2002 there were a few large studs that had kept pace with technology and had integrated traditional craft skills with objective measurement in a commonsense way. They will claim market share; those unable or unwilling to specify their product will fall by the wayside. The direction of the ram business will pass from the hands of sheep classers, legendary stud men and salesmen into the hands of consultants, information providers and analysts. My career spanned the best of the old stud world, and I'm glad of it. I suspect the new world will be a good deal more efficient, and that's cause for rejoicing. But it also might be a good deal more boring.

The changing of the guard in the 1970s: F.S.Falkiner & Sons was sold to News Limited. Alec Morrison, Prime Minister Bob Hawke and Ken Cowley inspect a top ram at a *Boonoke* field day *(F.S.Falkiner archives).*

CHAPTER TWENTY-ONE

A CLASS APART

They had succeeded in forming a sort of composite order of their own; in their estimation the perfection of human wisdom – but in reality a mongrel breed, in which may be found every shade between the real Australian Merino and the dried-up Leicester, mixed with myriads not fit to class as respectable goats.

PIONEER SHEEP CLASSER THOMAS SHAW ON AUSTRALIAN SQUATTERS AND THEIR SHEEP, CIRCA 1849

Between the 1850s and the 1960s the quality and style of Australian wool was enriched, to a great degree, by the activities of professional sheep classers.

The existence of the profession of sheep classing is made possible by the fecundity of sheep themselves. Every property has a finite capacity to run a certain number of sheep; it may carry a few more in a bounteous season and a few less in a poor one, but on average numbers must be kept at a certain level to avoid overstocking and preserve the natural environment. Each year the ewe flock will produce anywhere between 60 and 90 per cent of lambs, so with these fresh young sheep coming in at one end, something must be disposed of at the other end to keep numbers in equilibrium. One alternative is to dispose of the appropriate number of older sheep, but the entrepreneurial squatters of the 19th century were quick to see an alternative that, at the same time, would lift the productivity of their flocks through improved genetics.

Thus the profession of sheep classing came into being. The classer's role is to visit the property annually and physically inspect every one of the young ewes about to enter the breeding flock, and use his judgement and experience to reject (or 'cull') that proportion of the inferior producers as will enable flock numbers to be kept in equilibrium. This is usually between 25 per cent and 40 per cent, depending upon the number of lambs dropped. Then to select, from a suitable stud, rams appropriate to reaching the owner's breeding objectives, which will probably include increasing the weight of fleeces, improving wool quality, and increasing frame size and fertility.

The actual process of classing is these days carried out by forcing sheep into a long narrow pen (called a 'race') in single file so that the classer, walking alongside, can lean across to inspect the fleece and assess the animal's frame. The best races are raised so that sheep are presented directly in front of the classer. In earlier times, when labour was cheap and abundant, sheep were dragged by hand to present to the classer, who sat on a box. If yards work well and the flow of sheep is uninterrupted, about 3000 put through is a good day's work, but in the early days, as we shall see, the old classers achieved phenomenal tallies. The job requires intense concentration, particularly if the classer is required to sort the sheep into a number of lines. To watch a competent classer, with his eye in, his lines set and working at top speed, is a magnificent sight, and they make it look just so easy. When people try it themselves they usually find it is not as easy as it looks, and they become lost in indecision.

Prior to the 1850s Australian Merino sheep – by today's standards – were small of frame and yielded extremely light fleeces. But the quality of their wool was superb; the world had never seen anything to equal it. The early sheep expert John Ryrie Graham recalled these extraordinary wools: 'From about 1829 to 1840 the Australian wools had a character so uniform and fixed that an English wool-sorter or broker would with certainty select by touch alone from a bale of others a Botany Bay fleece, as they were then termed.'[1] Then, as Graham recalled: 'From about 1835 to 1840 a perfect mania for sheep-farming had taken possession of military officers, gentlemen of the mercantile navy, lawyers, clergymen, merchants and others with capital to invest, or credit to float a set of bills. Sheep and their wool were the all-absorbing topic of the period, and immense numbers of sheep and stations became the property of gentlemen who had not the slightest knowledge of sheep, stations, or their management. It is therefore not surprising that, from this time, the ruin of the Australian Merino commenced.' These new entrants to wool growing had hit upon the notion that they could increase their financial returns if they were able to place the fleece of the Merino on the massive carcases of the English Leicester or Lincoln sheep, and a fashion for cross-breeding swept the land. The superb qualities of the old Botany Bay fleeces were soon submerged in a welter of rough, shiny, slippery wool.

So concerned did the English wool processors and merchants become at the deterioration of the Australian clip that they sent a man out to investigate what was happening. Thomas Shaw arrived in Australia on his mission in 1843. He was described as a 'fiery little fellow, surprisingly fluent with words for a dour Yorkshireman'. Shaw later recalled: 'I came out with the fixed determination to get as speedily as possible amongst the sheep and to investigate and, if possible, to clear up the mystery of why the wool varied so much.'[2]

Shaw soon became the saviour of the Australian Merino, and gained a reputation among the squatters of the day as the best man to engage for improving a flock. Probably his greatest contribution to the Australian wool industry came in 1861, when he selected the foundation stock of the *Wanganella* stud for the Peppin family. These sheep became superbly adapted to the arid inland country, and to this day the descendants of the *Wanganella* bloodline remain the dominant influence in the Merino sheep flocks of Australia.

So busy did Thomas Shaw become that in 1852 he induced his 26-year-old wool-sorter son Jonathan to come to Australia from Leeds. This tall, taciturn and dignified Yorkshireman, who in appearance was not unlike Abraham Lincoln, was to consolidate and expand his father's practice, and between them they founded the profession of sheep classing. Jonathan Shaw handled 'all the flocks worth classing' in the 1870s and 1880s, and he was described at work: 'He was something of a martinet. He was as quick as lightning when he was classing; he had to be, he had so many flocks to handle, and there were no motor cars in those days. He would condemn with a nod a heavy-tipped sheep, or one whose lack of size or fluffy tip proclaimed an obvious cull. The others he would class with a quick, horizontal opening of the fleece into the top, flock or cull lines.'[3]

Jonathan doubled as head wool valuer for the Melbourne wool firm of Hastings Cunningham, later to become Australian Estates, and had to complete his classing run in the wool off-season, touring Victoria and the

Riverina 'in a very smart turn-out drawn by four well-groomed horses'. He had to turn away a number of prospective clients, including the important Victorian grazier George Russell, to whom he wrote in early 1869: 'I am sorry to say that I cannot undertake to do so. My time is fully engaged. I have not one day at my disposal up to the end of September and on October 1, I am bound to be back in Melbourne.'[4]

Sadly, Jonathan Shaw went broke on *Murtee* in the Wilcannia district after being caught up in the great droughts of the 1890s, and died a poor man in 1905, aged 80. No man since, with the possible exception of HC Pennefather, has exerted such influence over sheep breeding in Australia.

Hugh Claude Pennefather was the natural successor to Jonathan Shaw, and in fact was mentored by Shaw when he joined the Australian Estates wool department aged 19 in 1882. Like Shaw, Pennefather was a wool man first, who subsequently learned about sheep. With Shaw's encouragement, 'HCP' – as he was known – was appointed sheep classer for all the Australian Estates station properties in 1893, and became a full-time classer from 1916. By the 1920s he was handling 200,000 to 300,000 sheep and selecting 12,000 rams a year, which required him to travel about 64,000 kilometres by a combination of train and a 'light buggy and a slashing pair of ponies' – a punishing schedule in those days.

HCP was a fine looking man with fair wavy hair worn short, and a luxuriant moustache waxed at the ends. He was strong-willed and frequently clashed with the managers of stations he visited. His influence within Australian Estates was considerable, and Robinson, manager of *Terrick Terrick* in the 1920s, was relieved of his post after Mrs Robinson refused to allow Pennefather to stay in the homestead because of his abrupt manner. His influence on the leading studs and flocks of his day was almost as great as Shaw's in the 19th century, but his work did not win universal approval from breeders who competed with his customer studs. In 1906 FS Falkiner wrote to the head of Australian Estates in outrage: 'He [HC Pennefather] stated openly to a buyer, who took counsel from him with a view to purchasing our rams, that he would have to stay in the yards all night with them if he wanted to get the sheep he bought.'[5]

HCP passed on most of his work to two of his sons and retired to live at Berwick in Victoria, where he continued to ride with the local hunt. He died in 1951, aged 88. Pennefather had made sure his sons received a thorough grounding in the wool trade before they ventured into the classing yard. His eldest, Hugh Frank Pennefather, referred to as Frank, was born in 1894 and spent two seasons in the wool stores after leaving school, followed by another two years as a jackaroo on *Terrick Terrick*. After serving in the infantry during the First World War, where he won an MC at the Battle of Polygon Wood, he spent some time with a Bradford wool firm before returning to Australia to build up a practice under the watchful eye of his father. Eventually he was to take over the classing for the Australian Estates properties.

Frank Pennefather was a good-looking man, with the neatly trimmed moustache and crinkled hair of a 1930s matinee idol, and a little vain with it. His quiet and reserved manner, pronounced almost to the point of austerity, probably stemmed from an innate shyness, and he could appear 'stand-offish'. He had been at Scotch College in Melbourne with my father and had known me since childhood, but seldom called me anything other than 'Morrison', and on occasions when I was detailed off to drive him to town he invariably sat in the back seat in stony silence. He could be testy in the sheep yards. The Pennefathers favoured the old style of classing in which they sat on a box and the sheep were presented to a holder sitting on another box opposite. The holder gripped the animal by the jowls, while the classer pinioned its hips between his knees as he inspected the fleece. Once Pennefather overheard an unfortunate jackaroo working the forcing pen make a whispered aside to his companion: 'Put one up arse-first to see if he can tell the difference.' He immediately stopped work and refused to continue until the manager had made the boy give a public apology. After a long day's concentration seated on the classing box, Pennefather would pick up his cushion immediately the last sheep went through and stride out towards the homestead. He would often walk many kilometres before the manager had tidied up in the yards and picked him up. In the 1960s I handled a flock near Tullamore in New South Wales that, prior to my appointment, had been under Pennefather's control for many years. On my first visit I was handed a leather-covered book in which, for 40 years, he had succinctly recorded his comments on the sheep following annual classing. From time to time there would appear the curt instruction, 'You will buy a Wanganella ram next year to the value of no less than 250 guineas.'

Frank Pennefather was an intellectual who thought deeply about his profession and about industry issues, and never became caught up in the humbug that surrounded much of the stud sheep industry. Just before his

Classing of one-year-old rams in the *Wanganella* classing yards, 1964 *(F.S.Falkiner archives)*.

death in 1964 he wrote: 'There has been, and still is, too much "mumbo jumbo" talk about sheep breeding and sheep classing. I do not believe there is any such thing as instinctive genius; and the vague and meaningless chat about suitable matings, and "nicking", is to me so much nonsense.'[6] He was a consummate professional, and his discretion when dealing with competing studs in his stable was legendary. When asked about how the sheep were on another stud his reply invariably was, 'They could have been a little fatter'. He protected his commercial integrity, and never took on the role of salesman for the studs he classed, usually refusing to accept a commission from vendor studs when selecting rams on behalf of clients. He, like his father, travelled extensively in the course of his practice, but reduced much of the stress by travelling to a central point by train, and then having each client drive him half way to the next assignment.

Frank's younger brother Wilfred was an altogether different personality. With a forthright manner, his craggy features were not unlike those of the British character actor Sir Charles Aubery Smith, who often played the role of formidable army colonels. Wilfred certainly had a fearsome presence and an imperious manner. I once classed sheep for a man who had been his commanding officer in the Second World War, who told me of the fuss Wilfred – who had served in the First World War – made when his true age was discovered and he was taken off a troopship bound for the Middle East. I commented that the troops must have dreaded Wilf's disapproval – he replied, 'Yes, and by God so did I!'

Wilfred Pennefather's influence in the industry was delivered through handling a number of the important parent studs in South Australia, and a string of stations owned by the Clarke and Tait families in Queensland. Like his brother he was a complex man, but always referred to himself as 'just a simple shepherd'. The Pennefathers assumed the mantle of the Shaws and, like the Shaws, the flocks they controlled were instantly recognisable to astute sheep men. They were plain-bodied, often bare of wool down the legs, perhaps showed a few physical faults, but invariably carried beautiful long-stapled fleeces, soft to the touch and with the crimps so well defined as to appear etched-in mechanically.

Another group of sheep classers, who came from an entirely different background to the Shaws and Pennefathers, also had an enormous influence on the Australian sheep flock in the latter part of the 19th century and the first half of the 20th. These were men without any background in the wool trade, but who had learned the art of stockmanship and sheep breeding through a combination of a natural affinity with animals and acute observation of how sheep performed under tough environmental conditions. Their experience was gained in the dry interior country, so they tended to favour big-framed animals with robust constitutions and heavy fleeces of strong wool.

Prince among these was that towering figure of the Victorian era JC Darke. Darke had been associated with the pioneer sheep expert John Ryrie Graham, and began working with him in the 1880s. Darke was a big, strong-boned, rather fleshy man with coarse but handsome features. He was close to the Falkiner family, and Fred Knight, grandson of FS Falkiner, described him as 'a fine-looking man whose appearance would have attracted attention in any company. He had broad shoulders and a straight back, his head was well covered with grey hair and his strong face was adorned with black eyebrows and a flowing grey moustache. He had a booming baritone voice, and he and Otway often sang duets, "The Holy City" being their favourite. He had strong opinions, and if he took a dislike to anyone, he did not bother to conceal it.' Strong personalities like Falkiner and Darke were bound to clash from time to time, and Knight describes one occasion: 'One evening he and grandfather had an ungodly row. What it was about, I do not know, except that Grandfather had insulted him. They walked up and down the front verandah abusing one another, Darke vowing he would never darken the doors again. Gran trotted up and down behind them, trying to make the peace. It all blew over, and Darke darkened the door for many years afterwards.'[7] Falkiner was a great supporter of Darke, and in the 1900s gave him a £500 bonus in recognition of his work with the *Boonoke* sheep. In 1906 Falkiner wrote to a prospective client: 'To show you Mr Darke's ability as a sheep classer: five years ago he started classing at *Booabula* – prior to that Blackwood [the owner] could not sell any rams except to the butchers. Now he can dispose of all he can breed at high prices and this year swept the board at the Deniliquin Show.'

Darke had an unusual method of classing. He sat on a box between two benches set at about knee height, and 12 catchers slid sheep along the benches to present them to the classer, seated on their rumps and leaning away from him so he could view the side, belly, and part of the back. It must have been an efficient method, because at *Portland Downs*, near Longreach, Darke once classed 12,000 maiden ewes in a 12-hour day, a record that has never been beaten. Darke's successor WJ McCarthy described him as 'an exceptionally robust man and a prodigious worker'. He needed to be – *Isis Downs* at Blackall in Queensland, one of his flocks, frequently handled 40,000 to 50,000 maiden ewes. Fred Knight described Darke classing the top rams at *Boonoke*: 'It was a memorable experience. Each ram was skull dragged before him, and an amanuensis, pad in hand, made notes of his remarks, the last one usually being "Sound testicles and good bone".'

Darke died in 1917, and during his lifetime had done much to improve the frame, constitution and fleece weight of Australian sheep through his advocacy of 'commercial return first'. His influence was spread widely through the many thousands of rams that went out to the sheep runs of the nation from studs he controlled – *Boonoke* and its various satellite studs, *Uardry* and *Coonong*.

Another apostle for robust constitution and commercial return was John Blood Dowling, who claimed descent from the pirate Captain Blood. He liked sheep with big frames, and was fond of saying, 'Nobody can afford to put a Rolls Royce body on a Ford chassis'. Born in 1884 at *Retro* in southern Queensland, he gained early experience under his father Aubin Dowling at *Retro* and later at *Talgai West* on the Darling Downs. He made his name setting up a successful stud at *Coreena* in the Longreach district after taking over their classing in 1905. He classed the flocks on the string of stations belonging to the T.A.Field organisation, and studs at *Minnie Downs* and *Booabula*, and had a lifelong association with Otway Falkiner at *Boonoke*. By 1962 he claimed to have classed 3.5 million sheep on 103 different stations, and sold 60,000 rams. His biggest purchase in one day had been at *Murgha*, Deniliquin, where he selected 1000 flock rams for Clarke and Tait's Queensland stations, and 750 for Sir Robert Philp of *Thylungra* at Quilpie.

Dowling was extremely fond of collecting his commission on ram sales, which in those days was the shilling in the guinea (up until conversion to decimal currency all rams were sold in guineas – one pound and one shilling). He had no scruples about shifting an account to assure payment. In 1922 he had selected 119 flock rams at 6 guineas and eight studs at 25 guineas for *Alice Downs* at Blackall from *Barratta*. Manager CJC Cameron felt constrained to fire a letter off to his head office: 'Dowling made a special request for commission, which he intimated other breeders paid him straight away and prior to settlement and before delivery of the rams. As this was the only thing between us, and he proposed ringing up *Wanganella* to hear what they had, it looked as if we were likely to tumble between two stools, and miss a very good sale. I therefore closed the sale by promising that he would be paid the commission on his return to Sydney.'

In his youth Dowling had been a big husky man, but when I knew him in the 1950s he was a little gnome-like figure with a hearing aid and a piercing, high-pitched voice, so by this time was nicknamed 'Squeaker' or 'Deafy'. He was still physically tough. He had some affliction that caused fluid to gather in his abdominal cavity and, when working in the *Boonoke North* sheep yards, by 3.30 in the afternoon his little pot-belly threatened to burst his belt. A quick trip into Urana Hospital where, in his words, 'they tapped me and put a spigot in', saw him back in the bar drinking whiskey by 5.30. In my time as manager of *Boonoke North* he selected many big drafts of rams, and had a penchant for sending frightening and cryptic telegrams, such as, 'Ring me tonight is trouble', or, 'Field angry games up better stay on toes'. Another read: 'Rams delivered Winton 20 short believe eaten circus lions please revert re compensation'.

Dowling loved the Longreach district, and a story is told of one of his visits: 'Jack thought, talked and lived sheep. They were rarely out of his mind. At one Longreach show he was invited to the pictures by a client – he must have been a special client, as Jack would much prefer the Longreach club where he could talk and sell rams. The picture that night was a romantic one and came to a climax with a torrid love scene. With everyone very intent, Jack's penetrating voice could be heard round the theatre, "They tell me the rams are not working too well out here this year".'

JB Dowling died in Perth in November 1963. He was most influential in Queensland, where he helped create the popularity of Riverina-bred rams in that subtropical environment.

Other influential classers were the father and son team of Harold and Tony Watson. Both close and careful men, they kept a low profile, exercised a high degree of discretion, and just got about classing the huge flocks under their control. Between them they classed the biggest flocks of all – those owned by the giant pastoral companies Scottish Australian, New Zealand and Australian Land, and Australian Agricultural Company. Both were known among the station staff of these companies as 'Weekend Watson', because of the number of times they classed on weekends – it was always easier to schedule a company job for Sunday than it was that of a private individual. Harold Watson was born in 1889, and was overseer for Neilson Mills at *Uardry* in 1914. It was probably through the influence of the autocratic and dominating Mills – who, along with Otway Falkiner, bestrode the stud sheep industry of the 1920s and 1930s like a colossus – that Watson developed his practice. As well as the big company places the Watsons handled a number of studs in New South Wales and Queensland.

Sheep classers lived nomadic lives, often driving up to 80,000 kilometres a year visiting far-flung clients, and spent months at a time away from home. Much of their life was spent staying in other people's houses, where they were usually welcome guests. Those whose lives were shackled to one property saw something romantic in the classers' lifestyle, and they brought with them news of the outside world and gossip about the happenings and people of the pastoral industry. Those of a gregarious nature thrived on the life, but there were risks attached. The much-loved Ken Smith, one-time classer of *Boonoke*, *Portland Downs* and other leading studs, claimed never to have gone to bed sober in his adult life, and that anyone who did so was a fool. When staying at the Hotel Australia in Sydney it was said that his standard breakfast order was a plate of porridge and two double whiskeys – one to brace his nerves and the other to pour over the porridge. He frequently called at *Buckiinguy* to see my father, and always seemed to be driving a big American car with its boot well stocked with Scotch. I remember him as a small, neatly built man, nattily dressed in the old grazier style of well-cut sports coat, jodhpurs and leather leggings. He seldom spoke ill of any man, and would refer to someone as being 'a dear old bastard'.

Sheep classers were great gatherers of market intelligence for the studs with which they were associated – what properties were changing hands, who was unhappy with their current ram supplier, what was happening in a district, and so on. Fred Knight referred to JC Darke as 'a one-man CIA'. One of the great classers of the modern era, Tom Lilburne, of Hay, was said to know everything that was happening around the sheep industry in the Riverina, often before it happened.

Relationships between certain classers and competing studs were often hostile. Early in the 20th century FS Falkiner exchanged angry letters with James Kidd, head of the wool broker A.M.L.&F. Co., regarding JW Belcher, a noted classer retained by that company. Falkiner hated Belcher, and the feeling was amply reciprocated. Falkiner wrote: 'Belcher is a blackguard. After the infamous treatment I have received from Belcher I cannot allow him on any of our stations. Had I given him the classing of my sheep his opinion of them might have been different.'

Sheep classing could be a remunerative profession during the good times. A classer's revenue came from a combination of professional fees, retainers from studs, and commission on ram sales. In 1900, JC Darke's scale of fees was 20 shillings per hundred for stud sheep and 40 shillings per thousand for flock sheep in lots of 10,000 and over – a level that would yield $1000 to $1500 per day in current terms for classing alone. In the 1950s and early 1960s classing fees were generally a more restrained £30 to £50 a day, but established classers prepared to push their limits in regard to travel were generating incomes in excess of £10,000 a year (around $140,000 to $170,000 in current terms).

Many classers had a strictly professional attitude in regard to the collection of fees. AL Morrison, a famous classer during the first quarter of the 20th century, was once engaged to class a flock in western New South Wales. He took the train to Nyngan, where he hired a buggy and four smart horses. On arrival at the station sheep yards he was greeted by the owner, up from Melbourne for the occasion, the manager and overseer, all wearing ties, and about 15,000 ewes which had taken a week to muster. Morrison stepped briskly down from the buggy, looked scornfully over the rails at the waiting sheep, and asked, 'How much are they worth boss?' 'Five shillings,' was the reply. 'Then sell the bastards,' said Morrison, as he stepped back into the buggy, slapped the reins, and scooted off back to Nyngan. His full account was in the mail a week later.

By 1965 there were 65 professional sheep classers practising in Australia. But everything had changed, and never again would a few individuals influence the nation's sheep flocks, as had the Shaws, Pennefathers, Watsons and JC Darke. In modern times classers have tended to align themselves closely to one or two studs, in effect acting as a representative for those studs; few are truly independent professionals. Many operate in one locality or region, and fewer range across the whole industry, as did the early classers. The big pastoral companies were broken up during the 1970s, and their influential flocks dispersed. The smaller holdings that came about because of closer settlement often ran flocks too small to warrant the cost of classing, or were multi-enterprise farms on which sheep were merely pawns in the cropping program.

The sheep classers' control over the great parent Merino studs, that distributed rams across state and climatic boundaries and reached into sheep runs in every corner of Australia, had once enabled them to leverage their influence into thousands of flocks. From the 1970s these great establishments began to lose their dominance as the market for rams gradually fragmented, regionalised, then localised. Smaller studs that once stuck exclusively to one parent bloodline also began to shop around and mix their genetics.

From the 1960s sheep classers faced a new challenge from fast-developing technology that was able to measure objectively those sheep production traits that the classers estimated using their craft skills – wool quality, fleece weight and body weight. Some of the old hands had trouble dealing with this technology; many rejected it entirely in the belief that they and they alone possessed the keys to unlock some deeper genetic mystery. At a meeting of sheep classers held in Dubbo during 1965, convened to discuss use of the new technology, one of these old-timers severely embarrassed his more progressive colleagues when he stated his criteria for selecting a stud sire: 'The eye, being the most important item of all, should be observed with great care, as without it being absolutely right the male lacks all potential to become a successful sire. The slack, heavy eyelid is a marvellous indication of lack of vigour and lack of interest generally.'[8] For the following two days delegates walked around peering suspiciously at the pitch of one another's eyelids.

A lot of nonsense was spoken and written during the 1990s about alleged lack of productivity growth in Australian Merino sheep, compared to industries such as dairy and poultry, attributing this apparent lack of progress to the failure of sheep classers and stud sheep breeders to use new technology. In fact these people had done a tremendous job – between 1920 and current times Merino sheep have become bigger, easier to care for, are more fertile, and have the potential to cut heavier fleeces. During that same period the environment in which Merino sheep live has become more difficult and more complex. The challenge from parasites is greater; the fundamental skills of stockmanship have deteriorated. The quality of natural pastures has eroded, and sheep are run more intensively in more confined spaces – the Merino sheep is a selective grazer and thrives best when it has plenty of room to move. It could be said that the Merino was most at home on the big runs, and that it was doomed from the time they began to be split up.

It is likely that technology for measuring the productive capacity of sheep will improve, as will the ability to manipulate their genetics through science. But the judgement of skilled stockmen like sheep classers, using their craft skills, will always be needed at both ends of measured selection systems. First to remove from the group those sheep that, because of serious physical faults, should never be measured. Then, once the elite animals are isolated using measurements, to select those very few that should be given the most scope to influence the next generation.

The profession of sheep classing is not finished.

CHAPTER TWENTY-TWO

JB DOWLING GOES DROVING

Delivered the rams. All arrived in splendid order.

JB Dowling

In 1904 sheep classer JB Dowling, then a young man doing general station work, undertook an eventful droving trip by stock route and train, delivering rams from the western Darling Downs in Queensland to *Corella* station north of Bourke. His pay was 'one pound a week, find own tucker and gear'. Dowling described the area they traversed as: 'Truly wild country where I was, plenty bad cattle to handle.' He kept a diary of the trip, written up using an indelible pencil and a black-covered school exercise book.

October 10th

Left Mt Abundance in company with Arthur Chilcott, with 349 rams. Had dinner at the shed where we dropped one ram, and camped at a hut five miles further on.

October 13th

Took the rams another 12 miles and camped on Eurella.

October 14th

Arrived Muckadilla railway station that evening and trucked the rams 348. Camped in the ticket office.

October 15th

Caught goods train 6 am and arrived Cunnamulla 10 pm. Rams alright. Robert Linton then took charge.

October 16th

Shepherded rams on Cunnamulla Common and put them back in the yards at night. Linton went insane and locked up.

October 17th

Started away early in the morning with rams and a mate caught me up at dinner hour. Returned to town for lunch, Linton no better, called for letters none there. Returned to sheep and my mate went back to fetch on the things. I took the sheep on the next stage to Burrumbilla. Had to watch all night. All my things gone on ahead to the border.

October 18th

Went five miles, camped on the Warrego. Got the Emu Pad bad.

October 19th

Went another five miles. Passed two mobs of cattle. Had a swim and made first damper; failure, flour rotten.

Droving sheep on horseback in 1915 *(Image 158791, John Oxley Library, State Library of Qld)*.

October 20th

Went five miles. Mr and Mrs Linton in Cunnamulla to take their son (still mad) to Toowoomba.

October 23rd

Camped the night on Thurlagoona boundary. Rabbits running through the sheep all night, almost raining.

October 25th

Camped on the Tuen reserve and Thurlagoona boundary. Went on to the Tuen pub for rations. Got maggoty salt junk and some jam. Had a bob dinner there composed of rancid goat's butter and the salt junk. Awfully hungry and dirty no change since I left.

October 27th

Killed a rabbit and I rode back to the Tuen pub to get some onions and curry powder to stew him. He was 'Baal Budgerie'. Went a big stage.

October 28th

Went six miles ... Tinnenburra boundary rider came up for five minutes. Dog ate all the meat just after tea.

October 29th

Reached Owenewen boundary where a man from Corella took charge. Only one johnnycake all day. Had big tea of cheese etc from new man.

November 1st

Crossed the border. Passed through Barringun and camped for dinner in a give-and-take pocket. Took off all the clothes I'd been wearing from Cunnamulla and got into the Warrego with them with a bar of soap for two hours and had a general clean up. Went on four miles to night camp. My mare bucked the pack off.

November 4th

Passed out of Belalie and passed through Enngonia, where we got rations. The last cheese just sold as I got there. Camped on stony ridge. Ran the sheep over the Enngonia cemetery before I could see it in the Budda Bush scrub. You could see down the burrows to the coffins.

November 6th

Went a nine-mile stage for water. Got good water in a dam. Camped at 18-mile government tank. Ran out of tucker, and could only get a loaf and tin of 'Bullocky's Joy'.

November 7th

Did nine miles. Passed into Lissington and camped a mile from Corella. 'Bullocky's Joy' spilt over inside of one pack. Flour buggered.

November 10th

I went on to the head station and in afternoon came back with Mr Jones and delivered the rams. All arrived in splendid order.

CHAPTER TWENTY-THREE

THE GRAZIER/FARMER FAULTLINE

He knows nothing; and he thinks he knows everything.
That points clearly to a political career.

GEORGE BERNARD SHAW (1856–1950), *MAJOR BARBARA*, 1907

Woolgrower politics are like malarial fever – they come and go on a predictable cycle. The political temperature tends to rise as wool prices fall – so the disastrous prices of the 1930s brought with them an outbreak of political activism. A form and pattern was then established that was to be repeated whenever wool prices fell below current production costs during the 1960s, 1970s and 1990s.

First comes the price reduction, usually driven by the fundamentals of supply and demand, followed by the rediscovery of politics. Then a frantic search is mounted for someone to blame – a futile exercise when demand for wool is driven by powerful economic cycles and moves in line with economic activity in the countries where the wool consumers live. The next phase sees overheated growers galloping in many directions at once, advocating a variety of fantastic schemes and proposals. This is accompanied by lengthy and passionate debate about what should be done, rather than actually doing something. Governments are able to manipulate internal tensions – farmer versus grazier or Western Australia versus the rest – to deflect pressure from themselves. When all else fails they appoint a committee of inquiry, which either buys time until the issue fixes itself or changes industry structures. None of these actions stimulates market demand or lowers production costs, but they do make everyone feel better.

Wool politics are fought out between and within the various grower representative bodies, which are usually state-based. Nearly all wool political strife ultimately arises from the fundamental differences between herdsmen and ploughmen. Soon after closer settlement began in the 1920s a fault-line opened between graziers, who ran their sheep extensively on large areas and relied on wool for most of their revenue, and farmers, who integrated sheep with growing of crops and had dual revenue streams. The two groups held to different philosophies about business and life in general. Graziers mostly lived in dry areas, so owned large properties. They tended to be isolated from their neighbours, travelled long distances to town, and many had been to boarding school in the cities followed by a stint away jackarooing on another property. They considered themselves to be more worldly, more sophisticated, and having a wider view of issues than those who lived in the 'inside' country. The operations they ran were akin to those of the pastoral pioneers, so they espoused the pioneering virtues of independence, self-reliance and resilience. Generally they favoured vigorous free-enterprise systems, and had considerable political influence. Many became politicians, and thus unofficial representatives of the grazing sector. Governments soon realised they were a force to be reckoned with.

Because they saw themselves as the embodiment of the sterling virtues that helped build Australia, there was a whiff of arrogance about the grazing fraternity. Annie Proulx captured the prevailing attitude in *Close Range*, her book on the Wyoming rangeland and its people: 'There builds up in men who work livestock in big territory a kind of contempt for those who do not. [They] measured beauty and religion by what they rode through every day, and this encouraged their disdain for art and intellect. There was a sombre arrogance about them, a rigidity of attitude that said theirs was the only way.'

Farming families were seen as latecomers; they lived in closer-settled areas, relatively close to town, on smaller properties run in a more intensive way. So they usually possessed a greater sense of collective behaviour, and tended to favour systems of orderly marketing, price controls, statutory authorities and subsidies. The cultural gap between grazier and farmer was often evident in modes of dress. To attend a sheep sale the grazier was likely

Alec Morrison with Peter Wagstaff, of *Malvern Hills*, Blackall, Qld, at a sheep display in Longreach in 1988 *(F.S.Falkiner archives)*.

to put on a clean pair of moleskins, polish his R.M.Williams Craftsman boots, and wear a tie. The farmer dressed as if he had just stepped off his tractor, with work clothes and muddy boots, and was proud of it. Eric Hitchins, a West Australian woolgrower representative and himself a farmer, described his sector's attitude to graziers: 'Always they looked backwards, seeking to recapture the past that had vanished with the passing of Macarthur. They saw themselves as a race apart from other men. They designed a kingdom of their own that must forever remain splendidly isolated and always supremely self-sufficient. Their attitude towards responsible government was always that it must remain aloof from their affairs, except when called upon to yield to their bidding.'[1]

Originally the two groups formed their own representative bodies at both state and federal levels, but most were to merge during the 1960s and 1970s in an effort to increase their political power. Despite apparent unity, underlying philosophic tensions continued to bubble away, to surface during the brutal political civil wars of the 1970s and 1990s. But the balance of power had by then shifted away from the graziers, because more than 60 per cent of the Australian wool clip was being grown on mixed enterprise farms.

There has always been an inherent quality of savagery in the way woolgrowers fought their political battles – they treated them as if they are a violent recreational sport like football. People from the corporate world, themselves accustomed to occasional blood-lettings, were appalled by the ferocity with which growers tore at one another during the upheavals of the 1970s and 1990s. It could truly be said of those times that woolgrower politics were fought with an intensity that made a trade's hall faction fight look like a church picnic. The diminishing pool of talent from which grower representative appointments could be drawn made the situation worse in recent times. Prior to and for some time following the Second World War many large and successful woolgrowers with commercial business experience took an interest in industry politics, and made themselves available for appointment. Then, following the extended period of low prices that culminated in the crash of 1970–71, a good few lost interest and returned home to run their businesses. So a situation developed in which, with some notable exceptions, success in industry politics was seen to run in inverse proportion to how well the appointees ran their own properties. Growers became cynical when they observed that often the person who ran the most inefficient place in their district was representing them on their state organisation.

During the 1990s particularly – when battling growers were suffering financial stress, social dislocation and loss of community – there developed a feeling of contempt for all wool politicians. Woolgrowers became susceptible to the inducements of demagogues, snake-oil salesmen and radically conservative political parties. Membership of state farming organisations fell away, so that in some states only 30 per cent of growers belonged to the established bodies. During the extended crisis of the 1990s, the longest wool recession ever known, the Wool Council of Australia, which was supposed to be the growers' supreme representative body, declined into a quarrelsome rabble who deposed some of their worthy leaders and lost credibility among their constituents in Canberra. At the time I was the newly appointed chairman of an industry service provider named the Australian Wool Research and Promotion Organisation (AWRAP), that was supposed to report regularly to the

Wool Council as representative of its constituents. In a spirit of cooperation and goodwill the Managing Director Adrian Kloeden and I decided to brief the central committee of the Wool Council on some commercially sensitive issues. We made an error of judgement, because information leaked via the state affiliates within 24 hours. When challenged, one executive gave the reply, 'You did well; it usually takes only 12 hours.' At their time of greatest need woolgrowers suffered a failure of political leadership at the top. Because of low prices and lack of profitability during the 1990s and since, properties began to get bigger as the industry rationalised. As a producer becomes bigger in scale and gains commercial strength, he or she tends to do their own thing and loses interest in being represented by an elected industry body. Gradually state and federal grower organisations will come to represent the smaller producers, those with the most problems, and so will have to move to the left, as it were, and become more strident in their demands, particularly in seeking government support of various kinds. At the beginning of the 21st century woolgrowers were still a widely scattered community with varied problems, different beliefs, and divided loyalties. They still possessed a strong sense of individualism. So the concept of one supreme political representative body, inspired by universal goodwill and unity of purpose, is a utopian dream that is doomed to failure. As growers become fewer in number and bigger in their scale of operations, the commercial leadership of large and professional operators will replace the concept of political leadership. The change is sorely needed.

CHAPTER TWENTY-FOUR

SIR WILLIAM GUNN

The Gods are well pleased when they see
Great men contending with adversity.

ROBERT BURTON (1577–1640), *THE ANATOMY OF MELANCHOLY*

In August 1951 a small paragraph appeared in the rural papers announcing new three-year appointments to the Australian Wool Board. The names included a 37-year-old Queenslander, WA Gunn, nominated by the Australian Woolgrowers' Council. He replaced the belligerent JP Abbot, the former parliamentarian who had just led a successful but bitter battle against the setting up of a compulsory reserve price scheme for wool.

Gunn was born in 1914 at Goondiwindi in southern Queensland, and with brother Wally ran their expanding family properties during the 1940s. An accident that damaged one of his hands meant he was unable to cope with heavy physical work, so it was decided he should enter public life. He soon proved adept at the cut-and-thrust of industry politics, and at the age of only 29 he was on the executive committee of the Graziers' Association of South-West Queensland; from that base his rise was meteoric. By 1949, at 35, he was president of the Graziers' Association of Queensland, in 1951 came presidency of the Graziers' Federal Council and membership of the Wool Board. By 1952 he was a director of the Reserve Bank of Australia and member of the Meat Board in 1953. In 1958 he became chairman of the renamed Australian Wool Bureau and a member of the International Wool Secretariat (IWS); he was appointed chairman of the IWS in 1961. Along the way he collected a knighthood in 1961, and thus became Sir William. In nearly 50 years watching wool politics in action, and for a short period as a reluctant participant in its dangerous and foolish games, I saw no-one to equal Gunn in skill at harnessing its forces.

The IWS had been created in 1936 to promote the use of wool worldwide. Australia, South Africa and New Zealand funded its activities through a levy on their growers. Until Gunn came upon the scene it operated rather in the style of an exclusive gentleman's club. It was as if its members observed the world outside through the window of the comfortable smoking room, only talking to those it considered worthwhile, and avoiding the grubby lower end of town where the real business of wool was being done. During his term of office Gunn completely revolutionised the IWS. He lifted its profile, changed its line of attack, and turned it into the key player in the worldwide business of wool. He gave it personality and thrust. In the 1970s he forced through introduction of a controversial reserve price scheme in Australia, which was to stay in place for 20 years before crashing into dangerous rubble because those managing it lost the plot. As Eric Hitchins wrote in 1972: 'He became, and still remains, the central figure in wool affairs, influencing those events more than any other person'.[1]

Finally he was pushed off stage, kicking and protesting, in 1973 as being too controversial. He was only 59 at the time and still in possession of his formidable powers. He returned to Brisbane and commercial life, with mixed success. He had wide-ranging interests in prawn fishing, stock feed manufacturing, cattle breeding, motels and export of live animals. Sadly, his pastoral company went into receivership in 1977 and Sir William fell on hard times. It was around that time that my faith in human nature was restored by a call from the editor of a major metropolitan morning broadsheet, with whom a story had been filed that Sir William was living on hamburgers in an old caravan on the outskirts of Brisbane. The editor decided not to run the story, on the grounds that he did not wish to heap more shame upon someone who had done so much for Australia. Sir William stayed largely out of wool industry affairs, but from time to time he would sniff the scent of battle and, like some old Lion King, emerge from his lair to savage someone then return again, muttering and growling.

Sir William Gunn: a big presence in wool politics from the 1940s to the 1970s *(News Limited)*.

Gunn in his prime was a formidable physical presence. At more than 182 centimetres tall and 120 kilograms it was hard not to notice this big, shambling man who always looked unkempt. He cared little for dress or appearance, and no amount of expensive tailoring or high-quality woollen suiting could make him look tidy. It is quite possible this was all part of his act. His features, to put the very best upon it, were rugged. They were topped by a thatch of unruly hair, parted in the middle, which stood out like the bristles of a thick brush, and conferred on him an almost simian quality. His manner could at times be uncouth. The New York office of the IWS once flew Gunn down to meet the Ladies' Committee of the Texan woolgrowers. During a heavy night in New York Gunn had somehow managed to lose one of his shoes, and it was too late to buy a replacement. His companions on the flight south hit upon the idea of wrapping his shoeless foot in bandages and offering the excuse of an injury sustained in the course of duty. The good ladies of Texas, deeply impressed by the 'Sir William' tag, expected some slim and elegant English Lord with impeccable manners to step off the plane, so were bemused when introduced to a shambling, hung-over, rumpled and corpulent Aussie, who merely grunted at them.

But beneath an unprepossessing appearance and blunt manner lay keen intellectual capacity, a subtle intelligence, relentless determination and fierce energy. Truculent in defeat, triumphant in victory, Gunn never gave up until he got his way. His abundant political skills had been hardened in the crucible of woolgrower politics; he possessed the attributes of a great advocate, exercising shrewd judgement about when to bluster, when to threaten, when to flatter, when to intimidate, when to cajole. He applied these skills in feverish travel around world wool markets, pleading the interests of Australian woolgrowers. Gunn had the intrinsic power of those who hold a strong and unswerving vision. He appeared to care little about extraneous things, or what people thought of him. In the early 1990s when he was 78 and his physical powers were waning, I put him on a plane in Melbourne. Going up a busy stairway in the terminal his battered old briefcase split asunder and great bundles of paper flew everywhere. Scrabbling around among the feet of other travellers to pick them up would be enough to embarrass most people, but not Sir William, who blithely went on talking about the issue at hand.

Looking back, it seems as if Gunn spent all the 1960s on the hustings or in Canberra corridors, campaigning for additional promotion funds or dealing with various disputes in regard to the reserve price scheme. Gunn was a teak-tough and seasoned performer, firm in his beliefs and with a thick skin, so he was seldom thrown off track. But his resolve was sorely tested in 1963. At that time the woolgrower community was sharply divided between those who wanted to set up a reserve price and stockholding scheme, and those who believed the priority for restoring a flagging wool market was increased spending on consumer promotion. Gunn sat firmly in the promotion camp.

In July 1963 he was partway through a whistle-stop tour of the woolgrowing districts of the nation, campaigning for additional promotion funds, which meant increasing the tax levied on growers. He planned to speak to grower meetings in Naracoorte, Beaufort, Benalla, Sale, Albury, Warracknabeal and Hamilton in the space of a few days. When he arrived at Hamilton in the conservative Western District of Victoria on the cold night of July 29, 1963, there were 1400 people packed into the town hall and milling around outside. On the way in he had to shoulder his way through shouting demonstrators carrying placards reading

'Marketing before Promotion', and about one-third of the way into his speech hecklers at the back of the hall began to shout, 'What about marketing?', 'Where is our money going?' and 'How much are you getting out of it?' Then the missiles started to fly over the heads of the crowd. Gunn was hit many times, and soon his wool suit was besmeared and splattered. It was a night of high drama. The defining image of the wool industry in the 1960s became the newspaper picture of the massive figure of Gunn, his blue suit plastered with eggs and flour, speaking on through the tumult, the stage befouled by rubbish and his aides nervously puffing at cigarettes in the wings. Gunn never missed a beat. He just stepped back momentarily, then went on with his presentation as if nothing untoward was happening. As he walked off the stage he was reputed to have quipped, 'Well, at least I pushed Christine Keeler off the front page.' (Keeler was a young lady of ill-repute involved with a British Cabinet minister in a spy-sex-security scandal.) He was shrewd enough politically to know that such a violent demonstration, in one of the most conservative regions in Australia, would act in his favour. It is not beyond the bounds of possibility that he orchestrated it, or at least invited it. Such demonstrations were rare in the woolgrower community; nothing like it had ever happened before, and such behaviour was not to be repeated for another 35 years, when an explosive meeting at Goulburn in 1998 passed a no-confidence motion directed at the wool body I then chaired.

Gunn thrived on crises and controversy; they were as meat and drink to him. 'Many of his actions have been adversely criticised. To this he has become hardened, and apparently takes pride in being a controversial figure. He has admitted, with evident satisfaction, to a reputation for ruthlessness.'[2] He was often accused of being devious. According to Hitchins, he 'was constantly giving rise to the charge that he prefers to seek any objective by devious rather than direct means. Open conflict has been avoided and devious means adopted to placate and appease, meaningless assurances have been given and expedient measures taken.'[3] Gunn was untroubled by such accusations. Speaking at a testimonial dinner held in his home town of Goondiwindi in 1990, he lashed out: 'In 1965 I took a conscious decision that I would do everything I could to outwit the New South Wales Graziers' Association, and to ultimately get a reserve price. The Association was ultimately beaten at its own game of foul tactics. In 1965 many of the woolgrower leaders (including three members of the board nominated by the grower bodies) deserted me and tried to get rid of me. But they failed. I was determined that the tactics which I would use would be as dirty as those which had been used against me.'

Gunn's public speeches were forceful, delivered in a punchy, tough-guy style that went over well with woolgrowers, and he had the happy capacity to write as he spoke. There was also a certain degree of theatricality about his personality that made some people feel uncomfortable, but which he made the most of. He once arrived intentionally late at a forum being addressed by an opponent. He shambled noisily down the aisle during the high point of the speech, found a front-row seat after much shuffling, plonked himself down, and proceeded to open the morning broadsheet amid great rustling and settled himself behind it, giving out an occasional grunt. The speaker was completely upstaged.

Reactions to Gunn were seldom neutral. One Queensland grazier observed: 'Like many a woolgrower's representative before him, it has been said of Gunn that he was seldom right but never in doubt on the vexed question of a reserve price marketing scheme. He has never been in doubt although he has changed from one side to another and back again, being apparently both right and wrong whatever way you look at it!'[4] Hitchins granted him grudging praise: 'William Gunn has an undoubted place in any future wool history. However, the historian will find difficulty in reconciling the inexplicable actions of a particularly complex character, which cannot be judged by ordinary standards of behaviour. He possessed immense energy and capacity, and was fortunate that circumstances made it possible to adopt a mode of life that suited his temperament.'

Gunn was never a widely popular figure. Leader of the Country Party and Deputy Prime Minister Sir John McEwan saw Sir William as his successor, and attempted to back him into what was seen to be a safe seat in Queensland, but Gunn failed to gain preselection. It was a peculiar attribute of Australians of that era – particularly if they were country people – that they mistrusted anyone who appeared too clever, too cunning, too intellectual, or who thrust themselves forward. They were also suspect of people who were decisive, and Gunn was marked down on all scores.

Many have disagreed with Gunn's policies, others have quarrelled with his methods, and some were upset by his abrupt manner. But all must applaud his energy and his passion, and agree that he was the most commanding and influential personality in the wool industry during the period following the Second World War.

Sir William Gunn died in early 2003 in a retirement village at Goondiwindi aged 89, after a long illness. It was as well he remained unaware of the parlous state of the wool industry at the time – had some of the innovations he introduced in the 1960s and 1970s been properly managed it would have been in better shape.

CHAPTER TWENTY-FIVE

CAST OFF THE SHEEP'S BACK

The misfortunes of poverty carry with them nothing
Harder to bear than that it makes men ridiculous.

JUVENAL (AD 60–130)

Thanks largely to Bill Gunn's political skills, in 1970 Australian woolgrowers had finally crept in under the shelter of the reserve price scheme many had long hankered after. This followed nearly 20 years of vicious fighting and political brawling between the graziers, who generally tended to favour free markets, and farmers, who generally sought market intervention. It also followed two bitterly fought referendums of growers, both of which had rejected the plan, and an independent enquiry, which had found the concept of a reserve price to be economically unsound. In the face of disastrous wool prices, a succession of federal governments had been anxious to enact whatever supporting legislation the growers wanted, but had trouble sorting out what that was as the opposing grower factions slugged it out. Gunn cleverly took advantage of the prevailing sense of crisis to have a scheme introduced, probably illegally.

The Reserve Price Scheme that began operations in November 1970 was a buffer stock scheme – it bought and sold wool in the market to offset short-term shifts in supply and demand. The Australian Wool Marketing Commission, who managed the scheme, purchased and held wool that failed to reach the reserve price, and sold stockpiled wool when prices moved above the floor. The commission operated a flexible reserve decided from day to day, and their whole purpose was to dampen down volatility by cushioning sudden rises and falls. The government kept tight control over the commission's activities, and it was required to report fortnightly to the Minister for Primary Industry and the Treasurer.

Over the next 20 years the perceived objectives of the scheme were to change considerably, and those changes would eventually bring about its downfall. In 1974 the flexible reserve price was abandoned in favour of a fixed reserve, which guaranteed growers a minimum price for their wool, to be set annually. It was to prove a dangerous move with devastating consequences – the rigidity imposed by the fixed reserve, and the general expectation that the level of the reserve would never be lowered, was to be the industry's undoing under financial deregulation in the 1980s. The original purpose of the scheme was to protect growers, and their customers, against disastrously low prices and to insulate them against short-term fluctuations, not to guarantee them a living. But by 1983 the objective of obtaining the maximum sustainable price had crept quietly onto the agenda. That year the annual report of the Australian Wool Corporation, the organisation which was by now running the scheme, noted: 'The purpose of the Reserve Price Scheme is to provide greater stability and predictability in Australian currency terms, whilst also to achieve the maximum sustainable price.' Then in 1987 the government, which previously had retained the final say in setting the reserve price, handed that responsibility over to the Australian Wool Corporation, who were to set the price in consultation with the self-styled 'grower's parliament', a political lobbying group called the Wool Council. Because the Wool Council effectively controlled the appointment of directors to the Australian Wool Corporation, the setting of the reserve was being handed over to the growers, who were notorious for not being able to agree on anything unless it was how to get one up on the opposing faction. The process of setting the reserve price had been politicised.

The inevitable result was that greed took over. The reserve price was raised by more than 70 per cent in one year, from 508 cents per kilogram in 1986–87 to 870 cents for 1987–88. Events seemed to justify the prevailing mood of optimism, for in July 1988 everything looked rosy. The Wool Market Indicator stood at 1075 cents a kilogram, a comfortable 200 cents above the new floor. As usual, we all thought the good times would go on forever. At the F.S.Falkiner operation we were selling all the Merino rams we could breed, netting more than $1200 for a bale of wool, receiving an unheard of $100 per head for surplus sheep, and we were ploughing profits back into property improvements. Woolgrowers collectively had reached that dangerous state of euphoria in which we believed wool prices had reached a new plateau that could be sustained, and that the old boom and bust cycles were things of the past. Wool, in our view, was not expensive – prices had merely reached a level appropriate to the quality of the product, and that was no more than we deserved. Alas for false hopes, for we soon plunged into the abyss. Murphy's Law went into action – everything that could go wrong did go wrong. The things that went wrong were completely unpredictable, and well outside the control of the Australian Wool Corporation.

Under protection of the Reserve Price Scheme wool growing had become more profitable than other rural enterprises, and there was a rush into Merino sheep. There was also a series of bountiful seasons in the late 1980s, with the result that Australian wool production grew more than 55 per cent between 1980 and 1990, to reach the all-time record of 1100 million kilograms. There was a growing danger of oversupply. The fibre was also becoming expensive at processor and consumer levels. It was generally accepted that wool remained competitive with synthetic fibres at a price ratio of around 3:1. By November 1988, 21-micron wool in Europe was at a ratio of 5:1, and in the US it was 7:1. Processors began to substitute synthetics for wool in their product lines.

The rebuilding of processor's stocks at a time when the Wool Corporation's stocks had virtually disappeared largely drove the booming wool prices of 1988. This cyclic stocking and de-stocking had long been a feature of the wool market, and was the principal cause of price fluctuations. In normal circumstances prices would have fallen back when private stocks were replenished – but this adjustment was prevented because the reserve price had been ratcheted up behind the market.

Then China and Russia, who between them had been purchasing 20 per cent of the Australian clip, dropped out of the market. During 1988 and 1989 epoch-making events in both countries caused a 50 per cent reduction in their buying, with virtually no wool being bought by them late in that period. To make matters worse, China began cancelling forward orders. Many traders were caught holding stocks of Australian wool that they had bought speculatively, and they dumped these onto the market.

As if these problems were not bad enough, the Australian dollar began to firm relative to the currencies of wool customer nations. Jim Young, then managing director of the Australian Wool Corporation, recalled: 'Wool poured onto the market in the spring of 1989 as the world closed its order books and sat back to watch.' The corporation began buying heavily in support of the inflexible reserve price. In the spring of 1989 their purchases at auction represented nearly 60 per cent of offerings. In August they spent $60 million in one week alone. It couldn't go on. Support for the scheme began to fracture as the industry divided between those who wanted to tough it out at the 870 cents per kilogram floor and the free marketeers who wanted to abandon the whole scheme immediately. The atmosphere was heated, and it was easy to buy into the argument around the bush at that time. The issues came to a head at the Wool Council's annual conference held in southern Queensland, an event that became known as 'The Battle of Roma'. As Jim Young recalled: 'All the elements of the philosophical debate on economic reform, doctrinaire ideology about the role of government and private sectors were to collide just as the deepest crisis in the world textile industry savaged the wool industry.'

The government, which was guaranteeing the Wool Corporation's debt, started to run scared. The Labor Minister for Primary Industries at that time was John Kerin, a former economist who liked to refer to himself as 'nothing but a simple chook farmer at heart'. There was something of the Jesuit priest about John Kerin. With long iron-grey hair, dark eyebrows, prominent lower jaw, spare stooping frame, shambling walk and thoughtful manner, he carried with him the air of an intellectual and ascetic. He was popular in the bush, and many saw him as a decent country type; too nice to be in politics. But his diffident manner and gentle self-deprecatory sense of humour hid an iron determination that was shortly to be revealed. Kerin had taken independent advice on the situation in the wool market, and had become convinced that the reserve price had to be lowered. Any attempt to tinker with their beloved scheme was likely to be extremely unpopular among the growers so, hoping to limit political fallout, Kerin attempted to force the board of the Wool Corporation to make the decision. The board, led by newly appointed chairman Hugh Beggs, was equally as convinced that the 870 cents reserve must be retained at any cost. Beggs was a woolgrower from an 'establishment' family in the Western Districts of Victoria, and was a big man physically, with the general appearance of a good-natured John Bull. But he too was a quietly effective negotiator with an outstanding track record in the turbulent arena of wool politics, and he too had a steely resolve. Kerin cajoled, appealed to reason, argued, threatened, and put enormous pressure on the board, but they held firm.

At one stage he announced: 'I am not attracted to brinkmanship and will not provoke it. If provoked by others, however, my nerve will hold.' True to his word, on May 31, 1990 Kerin strode into Wool House in the Melbourne suburb of Parkville to confront the board and, using his reserve powers as minister, directed it to lower the reserve price to 700 cents a kilogram, which was supposed to be the long-term trend. This was probably the worst of all options because, on the assumption that if the government was prepared to lower the price once then they could do so again, the wool trade just sat on its hands and let things unravel. On February 11, 1991 the minister announced suspension of the scheme until the end of June. Then, as everyone expected, it was abandoned.

Closing stocks held by the Australian Wool Corporation at the end of June 1991 stood at 4.7 million bales, around a year's supply, and debt at $2.8 billion. The malign influence of both was to blight the affairs of woolgrowers for the rest of the decade. On February 26, 1991 wool sales were opened without reserve price support for the first time in 20 years. The Market Indicator price immediately fell nearly 40 per cent to 428 cents per kilogram. As Peter J Austin – who worked for the Australian Wool Corporation during the life of the Reserve Price Scheme – said in a retrospective: 'The RPS worked well for much of its life, but like a person free-falling from an airplane without a parachute, it was the bit at the end that caused all the damage.'

The Reserve Price Scheme had shielded growers from market volatility for more than 20 years. When it collapsed, woolgrowers were plunged into the darkest and most protracted period of depression they had experienced during the 20th century. To be sure, there had been tough times during the 1930s and 1960s, but none cut so deep or lasted so long as that of the 1990s – it was a grim battle of attrition. Production costs varied greatly between regions and even between properties. As a rule of thumb, during the 1990s growers needed a return of around $700 a bale just to break even. Because of the looming threat of the stockpile, the average price per bale was below cost of production for half the decade, and the other years were marginal at best, apart from a blip caused by pipeline restocking in 1994–95. In 1992–93 the average price was a disastrous $556 per bale.

The effect on many individuals was devastating. In April 1993 *The Land* columnist Peter Austin (not to be confused with Peter J Austin) reported: 'A substantial north-west NSW woolgrower told me last week that his wash-up for the sale of a 600 bale wool clip a few days earlier had been a net proceeds figure of $5.24 a head. That was before allowing shearing costs of $3 a head, and that of course was before allowing for the costs of running the sheep all the year – and this was for an outfit running close to the optimum level in terms of sheep per man employed.' Out in the tough country near Ivanhoe in New South Wales Graeme Rees, a hardy western grazier who knew how to handle the arid environment, revealed that net returns from his 6000 sheep would only cover just over half what he needed to break even. He reported that in 1989 his net return per bale after selling costs was $654, in 1990 it fell 30 per cent to $462, and currently he expected it to fall further to $262. Small growers and their families were particularly stressed. In January 1991 Anne and Rob Willis, both 34, who operated an 840-hectare farm near Inverell in New South Wales running 1500 sheep, told *The Land* they expected their wool returns to drop 50 per cent, and they knew they would get virtually nothing for their cull sheep. Rob Willis reported that to make ends meet he had taken on off-farm contract work, and he and a friend had recently been away in the St George and Collarenebri districts where they had crutched 80,000 sheep. He hoped to double that figure in the coming year, but lamented that it 'will put a lot of pressure on the family'. Anne Willis was looking at the prospect of running the farm single-handed in addition to caring for their four children: Amy, aged seven; Tom, five; Meg, three; and Kate, six months.

A few, amazingly, retained their sense of humour. In August 1991 Goodooga grazier Jamie Richmond wrote about the increase in shearing costs from 11.4 cents in 1921 to $3.11 a head in 1991. He recalled: 'I was born at *Mogila*, Goodooga, 70 years ago during the 1921 shearing. My father stopped the shearing, gave the shearers a tot of whiskey and drank my health. I mentioned this fact to the shearers a few years ago and one of them said: "Yes, and that's the last time a Richmond ever shouted for the shearers!" This oversight has since been rectified.'[1]

Inevitably, there were foreclosures and bankruptcies. During 1990–91 a bank took out a writ of possession and evicted a young couple from their

7000-hectare property near Tottenham in New South Wales. The bank had earlier encouraged them to expand their operations by buying more country and farming plant, but dry seasons in 1985 and 1986 had combined with high interest rates to push loan repayments into thousands of dollars a week. In the end they were unable to service their $1.7 million loan. Ominously, the bank advised that another 20 clients in the region were on their 'critical' list.

Hard times quickly translated into falling land prices, and there was a wave of concern even among those with manageable levels of debt as asset values eroded and banks became more exposed. The historic *Toganmain*, on the Murrumbidgee near Hay, had sold at the peak of the market in 1988 for the equivalent of $392 a sheep area. Resold as part of a larger aggregation in October 1992 it made only $194 a sheep area, and prices continued to fall through the decade. In March 1999 the nearby property *Oolambeyan* sold for $124 a sheep area. In the end, falling asset values did many a good turn, because banks were forced to rethink policy on their rural loan portfolios towards an emphasis on supporting farmers through the troubled times. In January 1991 five mortgagee sales in the Wimmera region of Victoria saw the properties

In the 1990s John Kerin took on growers over the Reserve Price Scheme *(News Limited)*.

passed in without a bid, while in the northern Mallee a property valued at $741 per hectare two years previously had fetched only $247 per hectare.

One of the high-profile crashes of the 1990s was the leading *Collinsville* Merino stud in South Australia. Its fall in 1991 symbolised both the excesses of the 1980s and the dark days of the 1990s. Prices paid for *Collinsville* rams had soared to values not seen since the 1920s. In each of the five years from 1985 to 1989 they achieved world records. In 1985 they had obtained $79,000, then in 1986 this was broken with $102,000 to an Argentinean syndicate. In 1987 their prices went first to $165,000 then to $215,000; 1988 saw flamboyant Queensland property developer Mike Gore pay $330,000. Then at the 1989 Adelaide ram sales a South African paid $360,000, and just six minutes later a local stud bid to $450,000 – a price that will never be surpassed.

The stud had been founded in the 1890s by the Collins family, who over the years developed a unique type of animal that combined the massive frames of the old South Australian bloodlines with the superior fleeces common to Peppin sheep. By the early 1980s the family company was suffering typical third-generation problems and, seeing an opportunity, the Riverina-based F.S.Falkiner studs, which I was then managing, prepared a confidential dossier on the *Collinsville* business and made an approach to the Collins family. This was done discretely through a third party, on the basis of opening discussions that might lead to an offer being made to the shareholders. Our approach was curtly rejected, so we decided to let the enquiry sit on the table in the expectation it would eventually attract the shareholders' attention. Frustrated by the lack of action, the consultant involved in compiling the dossier passed material on to the entrepreneurial Western Australian stud breeder Neil Garnett, who outstepped the Collins family in a series of negotiations carefully leaked to the rural press. He finally purchased the stud in 1984 for $10.5 million, financed by a $7.5 million loan from Elders Rural Finance.

Garnett was different to the normal run of stud men. Tall, good-looking, always well dressed in trendy style, he affected rather elaborate manners, and spoke in a deep, foggy voice. Many of the craggy old stud men felt distinctly uncomfortable in his presence – he was altogether too smooth for them. Garnett had a vision, strong business skills, and knew how to structure deals. He soon set about turning the conservative stud sheep industry inside out. He took the sheep-breeding program at *Collinsville* high-tech, making wide use of artificial insemination and embryo transplants. He attempted to create a new market for frozen ram semen, concentrating on export sales. He boosted prices for his elite rams to breathtaking levels through elaborate deals involving combinations of bank loans, guarantees of semen sales, and joint ventures. He was a consummate showman with an instinctive understanding of how publicity and public relations worked, and built a cachet around the stud that gave the perception that everything it did was 'leading edge'.

The first hint of trouble appeared in July 1991 when Steve Cooper, an astute investigative reporter for the *Weekly Times*, revealed that for the year ended June 30, 1990 the *Collinsville* operation had shown a loss of $5.8 million, and its debt had blown out to $23.8 million. He quoted 'industry sources' that claimed negotiations were underway to sell part of the stud to foreign interests – specifically, from Russia. In August Cooper wrote that *Collinsville's* debt, by now crossed over to the ailing State Bank of South Australia in mysterious circumstances, had gone out to more than $40 million. His investigations showed balance sheet assets exceeding $36 million, including $30 million for sheep and $5.2 million for ram semen stocks, but industry sources were questioning these values in the light of the slump in wool prices.

The axe fell in November 1991 when receivers were called in, and total debts of more than $50 million were revealed. The suspicions that many of us held about the true background to record-breaking ram prices seemed to be confirmed when documents revealed that, in many cases, purchasers took out joint bank loans with the stud around the time the purchases were made. Richard Nitschke, purchaser of the world record $450,000 ram in 1989, was quoted as saying: 'We got sick of it when we could not sell semen overseas. So we said: "You go your way and we will go ours," and we paid back our part of the loan.' When asked where the ram, 'Lustre 53', was, Nitschke responded: 'He's still around somewhere, but we are not collecting semen from him.'

A final poignant note to the whole dismal affair was sounded in a story in the *Weekly Times* of November 20: 'A lone security guard was the only movement at the *Collinsville* Merino stud – work had seemingly come to an abrupt halt just 24 hours after receivers moved in. The 50-strong staff had gone into hiding. Only the sheep escaped the tension which loomed over the property as they grazed unaware of the predicament the flock and the bloodline were in.'

The dramatic fall in wool prices triggered an inevitable, and necessary, shakeout in Australian sheep numbers and wool production. Numbers were shed as fast as they had been put on during the late 1980s, as financially stressed growers grew to hate sheep and turned to crops, cattle and goats. Across the decade sheep numbers fell by a massive 35.8 per cent, or 62.1 million, from 173.1 million in 1990 to around 111 million at the turn of the century. Wool production halved, from 1100 million kilograms greasy in 1991 to an estimated 550 million kilograms by 2002. At the close of the 20th century the adjustment had overshot, and the Australian sheep flock and wool production had fallen back to where they had been in the late 1940s. With the stockpile that had haunted the 1990s gone, it remained uncertain what the implications of still reducing Australian and world wool production would mean to the role of wool among the competing textile fibres.

CHAPTER TWENTY-SIX

BITTEN BY WOOL POLITICS

I long ago came to the conclusion that all life is six to five against.

DAMON RUNYON (1884–1946)

During 1996, as chairman of the industry body called the Australian Wool Research and Promotion Organisation (AWRAP), I was speaking at a woolgrower meeting in a hot and dusty hall in a desolate part of south-western Queensland. A dozen or so had turned up, many straight out of the paddock and still in working clothes. It was late afternoon, the day was warm, and their mood was quietly sombre. Some were dozing off, and a good few looked as if they would rather be someplace else. While speaking I was keeping a weather eye on one of the audience. He bore an unkempt look, with bushy beard and long, straggly, unwashed hair. He was restless, fidgety, and clearly angry. I felt sure he would prove a handful when question time came.

Right on cue, he jumped to his feet and shouted in a voice thick with rage: 'You need to get your bloody act together on selling the stockpile and stop dumping it on the market!' When I gently pointed out that as chairman of AWRAP I had nothing to do with the wool stockpile, which was the responsibility of a different body known as Wool International, he was undeterred: 'Well, that's all right, all you bastards need livening up anyway!'

He was confused, as were many others. Who could blame them? During the life of one woolgrower there had been a bewildering succession of organisations involved in some aspect of the industry. The Wool Bureau,

Wool Board, Wool Corporations Marks I and II, The Wool Marketing Corporation, Wool Commission, Wool Realisation Commission, Wool Research and Development Corporation, International Wool Secretariat, The Woolmark Company, and now Wool International and the Wool Research and Promotion Organisation. During the life of these organisations the astounding number of 111 directors had held office, and there had been platoons of chairmen. As US broadcaster Ed Murrow once said: 'Anyone who isn't confused doesn't really understand the situation.'

Following the dramatic collapse of the Reserve Price Scheme in 1991, and an inquiry headed by the former International Wool Secretariat managing director Sir William Vines, the government had appointed three separate bodies to clean up the wreckage; one to take charge of stockpile disposal, one to supervise promotion, and another to manage research and development. Each organisation was funded by a combination of government support and a tax levy on growers. The structure was clumsy, particularly in regard to separation of the promotion and research functions. Relationships between the bodies soon became poisoned. In some cases the managing directors were not talking to one another. There were constant demarcation disputes. Communications broke down. Wool prices meanwhile continued to languish as a result of the dark shadow of the stockpile and uncertainty about how it would be sold down.

In these troubled circumstances Simon Crean, recently reappointed as Primary Industries Minister in the new Keating Labor Government, felt yet another inquiry coming on. To head it he chose one of the Labor Party's favourite academics, Ross Garnaut, professor of economics at the Australian National University and former Ambassador to China. Garnaut's committee had a strong business stamp about it – the other members being Rod Price, aggressive chief executive of the building materials and petroleum company Pioneer International, and the creative founder of the Australian clothing group Country Road, Stephen Bennett. They were given only three months to complete the task and, unlike the Vines Committee that preceded it, would not conduct any public hearings. Their terms of reference were extremely broad, and no aspect of the wool industry was excluded from scrutiny.

Garnaut's recommendations resulted in yet another restructure and yet another set of confusing acronyms. The Australian Wool Research and Promotion Organisation (AWRAP) was to handle the functions its name implied. Stockpile disposal was put in the hands of Wool International

(WI), and The Australian Wool Exchange was to assume the various regulatory functions once handled by the old Wool Corporation.

One night late in 1993 I received a call from Mac Drysdale, a Queensland woolgrower who had been chairman of the now-defunct Wool Corporation. He advised he was to become chairman of AWRAP, and that I was to be appointed a director. Though I knew my name had been put forward, it came as a surprise. To be sure, I was flattered and excited, but at the same time felt cautious. I knew that the organisations needed substantial cultural change in the way they operated, but remained unconvinced they would be able to do so. Thus began a five-year involvement in public life, which was to end unhappily in 1998. I had been a vocal critic of marketing policy, believing that funds should be directed at branding apparel products made from Australian wool. So my first action on appointment was to drive to Melbourne and spend two days in Wool House studying all the consumer research, undertaken over many years, into the buying public's reaction to the Woolmark label, which was used to promote wool as a generic product. The reception I got there was frosty – the management had suffered through a great deal of organisational turmoil, and were understandably suspicious of a new boy on the block. To my surprise, I found that I was wrong – consumers in shops bought clothes for reasons of fashion, fit, feel, colour and price. They were little interested in what country the component fibre came from. It was a valuable lesson, for it taught me that woolgrowers that are critical of their organisations are often unaware of the complexity of the issues they grapple with. Mac Drysdale was to head up AWRAP, and the exuberant and energetic Dick Warburton, formerly Australian head of the chemical giant Dupont, was to chair Wool International. The directors were indeed an eclectic mix. Appointments included former Sportscraft chief and textile industry guru John Blood; David Farley, the dynamic young chief of Colly Farms Cotton; French-born Jean-Christophe Giraud, Australian head of wool processor Chargeurs; fashion designers Carla Zampatti and Jane Lamerton; banker Chris Roberts; Coca-Cola financial executive John Priest; and former New South Wales Minister for Agriculture and right-wing Labor strongman Jack Hallam. In *The Land* of December 9, 1993, wool reporter Brian Clancy commented: 'Cotton, synthetics, branded product, value-adding, fashion, finance, futures and politics. Like a recipe for a big cake, Primary Industry Minister Crean has taken all the ingredients he believes will get the wool industry back onto a healthy diet of prosperity. Mr Crean deserves credit for appointing people who have demonstrated successful marketing flair. In many instances the appointees have been mavericks. Roping their talents into Wool House should encourage a much-needed improvement in wool industry harmony.'

The organisations were to perform strongly over the next five years, but were unable to restore industry-wide harmony and were to hit troubled times due to forces outside their control. Wool prices continued to languish and there was a constant muttering of unrest among the growers. The wool stockpile was the dominating influence during the 1990s – it was interesting that market confidence was restored immediately the final bale was sold in 2001. Business people understand the dictum 'your first loss is your best loss', and when they have an overhang of slow-moving stock they get rid of it as fast as possible. But the government, sensitive to problems in regional electorates, dithered, vacillated, listened to those who shouted loudest and couldn't refrain from interfering with orderly and speedy disposal – much to the frustration of the stockpile managers. At first stockpile reductions were based on a sensible flexible-release program, which enabled the managers to take advantage of market cycles. Then it was changed to a debt reduction schedule, then to an inflexible fixed-release schedule under which a certain number of bales had to be sold each quarter. When the wool market failed to respond the government yielded to political pressure – coming mainly from Western Australia – and changed the schedule yet again. The customers, many of whom had been severely burnt when the reserve price had been abandoned, looked on cynically at the government's dabbling, and only bought wool on a hand-to-mouth basis. The market stayed marooned.

During 1996 AWRAP chairman Mac Drysdale – disillusioned by the constant political bickering and sniping so beloved of woolgrower groups, and anxious to try his hand at commercial life on public company boards – decided to resign. I had been an AWRAP director since its inception, and Drysdale sounded me out as a potential successor. I was a reluctant candidate. Aged 61 at the time, I was heavily committed as chief executive of the Riverina-based F.S.Falkiner stud sheep empire, which was then a subsidiary of the media group News Limited. So I was concerned about the calls on my time. The chairmanship of AWRAP also involved chairmanship of the International Wool Secretariat (IWS) and heavy commitment to overseas travel. I had contracted, and apparently beaten, kidney cancer seven years earlier, and that too was in the back of my mind. I also felt repugnance for industry politics that almost amounted to contempt, particularly the back scratching and deal making involved. Perhaps those attitudes had been acquired during my youth – my mother once told me I could do no wrong in her eyes, and that the only circumstances under which she would disown me were if I committed murder or became a politician.

My final acceptance was motivated by a number of factors. To be sure, there was a hunger for achievement. There was also the excitement of coming to grips with the strategic issues, learning new things and mixing with people at the peak of the industry worldwide. Then there was the old-fashioned concept of duty. I was the son of a First World War digger, had spent my childhood enveloped by the Second World War, and attended a church school that drummed into us the importance of service and the Protestant work ethic. My generation absorbed many Victorian-era values. Finally my chairman, News Limited's Australian head Ken Cowley, encouraged me in his gentle way to take on the challenge, so I eventually accepted nomination with mixed feelings. I advised Primary Industries and Energy Minister John Anderson that I would serve for one three-year term only, that I would bring along a successor, and would push hard to implement the reforms begun under Drysdale.

My appointment was not popular among the woolgrowers. I was the first chairman in many years not to have been drawn from the ranks of a grower representative organisation, so I was not beholden to them in any way. Despite a deep and passionate involvement with the industry for 40 years, some did not regard me as a 'real' woolgrower because I was a company executive and not an owner–operator. Others were irritated by my rather flamboyant speaking style and a tendency to preach to people, and by my use of stagecraft and props like funny hats to illustrate points. They thought it was beneath the dignity of the position. However, I did not take the job on to win a popularity contest, and that was just as well.

AWRAP, its subsidiary The Woolmark Company (formerly IWS), and Wool International achieved much during their short lives. They could have achieved a great deal more if given a little extra time on the field. They were smashed on the shoals of industry politics, and replaced by yet another set of organisations. It is customary in business and politics for any new organisation to denigrate the performance of its predecessor so as to enhance its own. The wool industry is no exception, and a set of myths were created about the alleged lack of efficiency and effectiveness of WI, AWRAP and The Woolmark Company. Like all myths they were difficult to dispel once they gained wide circulation. Nevertheless they are myths, as anyone who worked closely with the organisations knows.

The changes created by AWRAP were quite radical, in fact in less than five years it completely reinvented the organisations and the way they operated. Yet many in the woolgrower community, preoccupied with the short term, just didn't notice what was happening. The International Wool Secretariat (IWS) was Australianised, corporatised, rebadged to become The Woolmark Company (WMC), and its operations integrated with AWRAP under one management. Activities were commercialised, including shifting Woolmark licensing to a fee-paying basis. External revenues of $70 million were built up to supplement woolgrower levy funds. Activities were outsourced and staff reduced from 1200 to fewer than 400. A comprehensive strategic plan for research and development was written. Research projects were reduced from a wasteful 500 to a targeted 80. A portfolio management structure was introduced, as was a new system for analysing and approving projects for funding. Regional advisory boards, involving chief executives from leading retail and processing firms, were appointed in Asia, Europe and North America. Under energetic managing director Adrian Kloeden the organisations began to effectively measure their own performance for the first time in their long history. High-order measurement of organisational performance was introduced, and performance-driven remuneration packages became

A gathering of 'wool tribes': the Goulburn meeting that brought down Alec Morrison's AWRAP board *(News Limited)*.

the order of the day. Even the board happily submitted itself to measurement of group and individual performance using an external consultant and critical assessment by the senior management group. Performance measurement of that kind was a very new concept then, and AWRAP was leading the pack.

However, woolgrowers' expectations of what AWRAP and WMC could do for them were too high. The organisations could lift demand for wool above what it would have been had they not been there, and consistently did so. But they could not control long-term cycles, strategic shifts in consumer spending away from apparel, or the health of national economies. They could not influence the price of wool in the short term; this was controlled by stocking and de-stocking cycles along the long wool pipeline. The kick upwards in wool prices to a market indicator of 840 cents/kg in 1994 was due to pipeline restocking, and proved short lived; by April 1996 it was back to 550 cents/kg, and in October 1998 plumbed the depths at 465 cents/kg. Above all other things, the organisations were unable to make the stockpile disappear without pain. As we have seen, continued low prices in the 1990s caused untold misery for woolgrowers and their families, and this in turn triggered an emotional atmosphere and renewed political activity. In this environment growers tended to seek out someone to blame for their troubles, and those in positions of authority became easy targets. At meeting after meeting passionate growers would stand up to pour out their genuine tales of woe, and attack AWRAP, WMC or the government for not doing enough to fix their problems. They were usually greeted by approving nods and warm applause. It was to no avail to point out the limitations of collective activity, or that the solution often lay in their own hands. Their reaction was not entirely rational, but understandable in the circumstances.

It was in these conditions that AWRAP managing director Adrian Kloeden and I decided we would try to steer the organisation well clear of industry politics and concentrate exclusively on commercial business agendas. AWRAP had become something of a punching bag, a passive target for anyone in the industry who had a problem. Some of those swinging punches were from organisations that were also applicants for funding from AWRAP, so we decided these people should clearly understand that we were capable of hitting back. We chose targets carefully and proceeded to rough up those people who tried to involve the organisation politically. Of course we made enemies along the way.

In hindsight, the attempt to stay clear of politics, though it suited my personal agenda, was a tactical error. One of our problems was that the peak grower body Wool Council of Australia, which had traditionally fought the political battles, had lost its authority. Driven by interstate rivalry and internal jealousies, many growers felt that it no longer represented them, and they were right, so its effectiveness in Canberra was diminished and ministers and bureaucrats began to listen to alternative views from dissident groups.

As the political atmosphere heated up, Kloeden and I had to spend more and more time dousing political bushfires, which became a distraction from running the business. As an example, we had received an application for research funds for a project using wool as a component in building material. Though on the surface the project looked attractive as a potential volume outlet for wool, in reality the applicant was seeking a large sum to provide working capital to start up his business. This was not research, so was outside AWRAP's funding guidelines and, quite properly, the application was knocked back. We introduced the applicant to firms that provided venture capital but, undeterred, he trotted the proposal around wool-growing districts and his cause was taken up by some of the grower organisations. Local members of parliament then put pressure on various ministers, and as a result we had to spend too much valuable time defending our quite proper decision not to provide funding.

In the prevailing climate it was clear that the organisations were becoming increasingly ungovernable because of mounting political pressures, and there was growing risk that our commercial agendas would be drowned in a search for political solutions. On a personal level, though I was still enjoying the challenge of helping manage the business, the politics had become very trying and it was getting so that I wondered what I was doing there. So it came about that one weekend in late 1997, Kloeden and Setha Lingam, a clever and impatient young strategist, and myself met at *Boonoke*. We sat on the verandah, drank tea and speculated about the future. From that discussion emerged a proposal to turn AWRAP from a statutory body, set up by Act of Parliament and responsible to the government, into a shareholder company operating under corporate law and owned by the woolgrowers. The shareholder company structure would make it easier to generate external sources of funding and revenue, and insulate the company from many of the destructive political pressures. We developed the proposal to outline form, and I trotted it along to Primary Industries and Energy Minister John Anderson for his reaction. He showed little interest, no doubt because a federal election was looming and he was anxious to avoid provoking further trouble in the already vexed wool industry.

But I found the proposal exciting, and also saw handing their company over to the growers as providing a nice point at which I could exit public life. I fully expected the various grower bodies to be equally excited, and to grab the idea with both hands, so was disappointed to find their reaction was cautious, even suspicious. I made a presentation to the Wool Council, the peak grower body, at a meeting in Perth, and was astonished to be met with stony silence. No-one seemed to understand what I was talking about, and there was not a single question. Then I realised that they were thinking ahead of me. They stood to lose much of their power as lobbyists, even the reason for their existence, if growers became shareholders in AWRAP as a company – shareholders would have little need for an intermediary to stand between them and the company they owned.

By this time political tensions in the industry were coming to boiling point in the lead-up to the 1998 annual general meeting of AWRAP at which the board was to be deposed. We considered bringing the proposal forward in that debate, but in the end thought it prudent to keep it under wraps to save it from becoming soiled in that ugly and overheated climate. Perhaps we should have gone ahead – three years later the industry service providers were converted into shareholder models, but only at great expense and after further turmoil.

A new lobbying organisation had appeared upon the scene in the middle 1990s. A breakaway group of dissident woolgrowers, which felt that the Wool Council was not adequately representing their interests, formed the Australian Wool Growers Association (AWGA). Their main platform was to return AWRAP to the type of grower control and domination that caused the crash of the Reserve Price Scheme in 1991. It was their view that directors should be woolgrowers appointed by popular vote, without any sifting process, in the hope that an exercise in popular democracy would solve a commercial problem. Other than freezing stockpile sales, which would have had disastrous consequences, they had few suggestions about how to adjust the overhang of stocks and lack of demand that were depressing prices. But, in desperation, woolgrowers were clutching at anything, and the AWGA found a ready audience.

The AWGA and other dissident growers made attempts to disrupt the AWRAP annual general meetings in 1996 and 1997. These included heckling, challenges to formal motions, persistent and petty points of order, obtuse and preposterous legal arguments, and long harangues from zealots, mostly obsessed with single issues of little importance. They were largely ineffective, because we had ample warning and were able to moderate their effects, but they were sufficiently visible to be embarrassing for the industry. At the 1997 meeting we had as guests two senior US executives from the giant consumer marketing company Procter and Gamble, with whom we were negotiating a deal to promote Woolmark and woollen apparel use. They were astonished by the pettiness of it all, and must have wondered what they had got themselves into.

But the dissident growers learned their lessons, bided their time, and set themselves to take advantage of a weak point in the AWRAP's defences. The Act of Parliament under which the organisation operated provided that in the event of a vote of no confidence being passed at an annual general meeting by those present and voting by proxy, the minister was obliged to sack the board. But the Act set up no hurdle in respect to the percentage of growers who were required to vote, and this created the opportunity for a minority of growers to unseat the board.

The AWGA and their supporters soon found an unexpected ally. In 1994 the AWRAP had appointed three Zone Advisory Committees (ZACs) of woolgrowers to assist with on-farm research. Members were paid sitting fees and travelling expenses, and were given the task of advising management on research needs, feasibility of projects, and adoption of new technology. Theirs was purely an advisory and consultative role – they were not appointed to take part in management decisions. The ZACs were a strange assortment of successful woolgrowers, frustrated politicians and scientifically minded incompetents who could not run their own farms properly, and sooner or later they were going to cause trouble. One faction became ambitious, and attempted to widen the role and scope of their activities to embrace post-farm research and even promotion. They sought direct access to the board, and increasingly saw themselves as taking a hands-on management role. This went far beyond their brief, and their behaviour revealed a fundamental misunderstanding of the respective roles of board, management and advisory groups. Had the board yielded to their requests its own authority would have been eroded, and executive managers in the research and development section put it in an untenable position. Another grower director, Rob Gillam from Western Australia, and I interviewed the ZAC chairmen and pointed out the error of their ways quite bluntly. Alas, they took it badly. The dissident faction couldn't help themselves. They became truculent and began to engage in aggressive political activity to achieve their aims, including deputations to the minister, the Wool Council, and state farmer bodies, and the leaking of confidential information. At one stage I was forced to threaten legal action when a clearly defamatory letter appeared in a Western Australian newspaper. All this was wasted time that should have been spent on more important issues, but such is life in the murky world of wool politics.

As could be expected in a period of low wool prices, calls for a wide-ranging inquiry began. Soon a loose coalition of dissidents was formed, with the one common aim – to unseat the board through a vote of no confidence, as the first step leading to another inquiry and the major restructuring of industry organisations. The coalition comprised the AWGA, the dissident ZAC faction, and former directors of AWRAP and the old Wool Corporation who had failed to be reappointed for a variety of reasons. There were also people seeking jobs or board positions, those philosophically opposed to collection of compulsory tax levies, embittered former employees, and the usual gaggle of power seekers. There was, as always in wool politics, a small group who wore their anger like a badge, and loved nothing better than conflict and public strife, because it relieved the banality of life out on the farms and stations, and used what they felt was their under-utilised intellectual capacity. There were even Liberal party backbenchers who hoped disruption might prise loose the National Party's grip on the Primary Industries portfolio. The rebels found plenty of support among struggling woolgrowers, many of whom believed an organisational shake up would somehow restore them to prosperity.

The months leading up to the 1998 annual general meeting were fraught and frantic with activity. The grower directors – Rob Gillam, who was a big pastoral woolgrower from Western Australia, and Barry Walker, a successful fine-wool producer from Yass in New South Wales – and myself stumped the country addressing grower meetings, most of which were hostile. We appeared in sheep yards, in corrugated iron sheds, in shire halls. In one week I spoke to more than 1000 growers at meetings in Inverell, Young and Jerilderie. But it was becoming clear that, with grower's businesses in a distressed condition, the forces opposing the board were gathering momentum. One Sunday at this time

I received a phone call at home from a woolgrower from the south-east of South Australia, one of the discontented ZAC members purporting to be speaking on behalf of the dissident coalition. He outlined the details of the plot being hatched, said that many of the plotters appreciated my work for the industry over the years, and urged me to resign as AWRAP Chairman to avoid being damaged by the explosion that was coming. It was an outrageous suggestion, and I told him so. My resignation would imply that the board had acted improperly or incompetently, which was clearly not the case. It would have been the 'dingo' option.

The AWRAP board's defence of its performance was seriously weakened by a *Four Corners* program that appeared on ABC television during the lead-up to the annual general meeting. The ABC's interest had been aroused by a public relations firm employed by some of the dissidents and, in the true traditions of investigative journalism, the show was flawed, to the extent it appeared producers reached their conclusion first and then constructed evidence to suit. They trotted out a research project of doubtful worth that had in fact been abandoned years before and a negative audit of research programs that had been commissioned internally and the problems subsequently acted on. Opponents of the organisation were given free play, and supporters unfairly edited. We had been invited to appear, but were advised against doing so by our media consultants, who had woken up to the fact that it was a hatchet job and that we were on a hiding to nothing.

The critical annual general meeting was held at Goulburn in the Southern Tablelands of New South Wales on Monday, November 30, 1998. As the various woolgrower tribes gathered at the meeting venue the air crackled with anger and tension. Passions were stoked up further when the original venue was not able to hold the more than 1500 who turned up, and the meeting had to be switched to another location. Also when a good many who were members of family companies or partnerships found – when they went to register for the meeting – that they had failed to nominate who was to vote for them, as was required by law and had been clearly advised in writing, they believed, wrongly, they were being deliberately disenfranchised.

There were a number of threats and rumours circulating, and the AWRAP had received warnings of a nature serious enough to warrant engaging a professional security firm when organising the meeting. Meeting venues were subjected to security checks, cleared by sniffer dogs, and the risk of some kind of physical confrontation was very real. Our worst fears seemed to be confirmed when, at the adjournment necessary to switch venues, an angry grower with arms waving stepped up onto the stage and began ranting at me. Fortunately he settled down. On the advice of our security firm, the young men and women who acted as microphone attendants in the body of the meeting hall were all security people trained in martial arts. I had fears that some grizzled woolgrower who refused to relinquish the microphone might be subjected to a sharp knee in the groin or swift karate chop behind the ear. Fortunately there were no incidents, but at one stage it got very close.

Security precautions included concealing the whereabouts of directors by use of dummy motel bookings and the assignment of a 'minder' to Managing Director Adrian Kloeden and myself. This former police detective was a formidable physical presence, and was briefed to shadow us at a discrete distance wherever we moved, even to the toilet. When at one stage the meeting was adjourned for the counting of a ballot, I joined hundreds of others moving to take a breath of fresh air outside. As the crowd shuffled through a doorway I was suddenly gripped from behind in a headlock, and a voice growled: 'I've got you at last you bloody bastard!' My heart sank, not only in fear for my own physical safety but in fear for the image of the industry should physical violence be reported in the press. I twisted round and firmly planted an elbow in the midriff of my assailant, only to find it was an old rugby opponent, whom I had not seen for 20 years, having an ill-timed joke. Fortunately I was able to signal the minder – who was moving in, hands circling, palms open ready for a knockout blow – that everything was okay, and a nasty incident was avoided. But only just!

The meeting, though tense and nerve-racking, went off with nothing more than some heckling, booing and foot stamping from a rowdy element, who seemed to enjoy themselves. It was all rather childish and reminiscent of a noisy school concert. But it was an altogether unpleasant experience for the directors, sitting on the stage like Aunt Sally and feeling the palpable hate coming from some sections of the crowd. Predictably, the no-confidence motion was passed, with 76 per cent of those voting in person and by proxy supporting it. The majority of large growers who voted by proxy supported the board, but this was insufficient to counteract those from smaller growers who had been skilfully gathered by the dissident group. But in the end fewer than 10 per cent of growers had registered to vote.

My personal feelings were of disappointment, tinged with a little anger. The board was amply skilled and had worked energetically at the process of reform, and they didn't deserve what had been done to them. The chairman's job was supposed to be a part-time role, but over two and a half years I had devoted 60 per cent of my time to the organisation.

But I also felt relieved – dealing with the business issues had been fun, but during the preceding 12 months the political warfare had been debilitating and I was looking forward to relinquishing the job in six months when the term of my appointment ended.

The following day Adrian Kloeden and I called on new Primary Industries and Energy Minister Mark Vaile in Canberra and briefed him on what was required under the Act. He then formally ended the appointment of each of the AWRAP directors, with the exception of the managing director, who was required to stay on until a new board was formed.

It was a bitter blow for the directors, who were passionate in their commitment to wool and to the organisation. Vaile attempted to soften the impact, and perhaps captured the mood of the time in his formal letter to me: 'I have been given to understand that even those that moved the motion of no confidence indicated it was not a personal condemnation of individual members. Rather, it was a symbol of the frustration many woolgrowers have with the restrictions imposed by the Act and the current wool market situation in general. I understand that this may be a cold comfort to you at a personal level, but please recognise that I believe the no-confidence motion was not directed at you personally. I know that during your service AWRAP did not resile from hard decisions as it streamlined both AWRAP and The Woolmark Company's operations … your contribution to these deliberations is recognised and appreciated … On a personal note, both my predecessor, the Hon John Anderson MP, and I would like to express our thanks for the timely and useful advice you have provided to the Government as the Chair of AWRAP.'

Following the dramatic events of November 1998 another inquiry into wool industry structures was inevitable. The government needed to clear the air and find a way to exit their involvement in the industry as soon as possible – there was only likely to be more political vexation if they stayed around. They acted quickly. In December 1998 former Liberal minister Ian McLachlan was appointed to head up a Wool Task Force. McLachlan came from one of Australia's old established pastoral families; a strange mixture of patrician and self-proclaimed hard man, he had made his name as a strike-breaker and scourge of trade unions during his term as president of the National Farmers' Federation. He had entered politics with wide support from the press, and many saw him as a future prime minister. Though he had made it to the front bench, he always seemed uncomfortable in politics – he somehow fell short of expectations, and many felt he had left the arena with his full potential unrealised. The taskforce was given broad terms of reference, covering the competitiveness of wool as a textile fibre, the profitability of wool businesses, ways of funding research, and existing institutional arrangements. They received 650 submissions. Like those heard by other inquiries before, many of these were serious and well intentioned. But, like an army regimental aid post, it also heard from the usual gang of cranks, crackpots, eccentrics, malcontents, malingerers, angry minorities and job seekers. Given the short space of time available the report they handed down on June 30, 1999 was a credit to the members, for they came up with many practical and commonsense recommendations.

Appointment of the task force was a political response to a political situation. Their report had a salutary effect – it went towards partly cleansing the body and spirit of an ailing industry. For a while at least the level of cooperation between hostile woolgrower factions was unprecedented in the industry's troubled political history. Western Australians sat peaceably with eastern staters, farmers with graziers, state farmer bodies with rebel groups. However, the peace was not to last. By the end of 2002 the body that replaced the AWRAP was becoming caught up in political warfare.

The demise of the AWRAP board was to trigger the end of collective funding by woolgrowers of generic promotion of wool that had been in place since the 1930s. The Woolmark Company was to be stripped of its effectiveness in the marketplace, and by 2002 the long-term implications of this remained unseen. One of the more contentious aspects of the report was that it paid little attention to the final consumers of wool, and the need to influence them to favour wool on retail shelves: 'Levy-funded generic advertising at woolgrower expense should cease forthwith. The word "Promotion" should disappear from the lexicon of the new organisation.'

The government had privatised the organisation handling the disposal of the wool stockpile in July 1999, so at the turn of the 20th century they had no direct involvement in the wool industry for the first time in 80 years. Given its contentious nature, who could blame them?

CHAPTER TWENTY-SEVEN

INNOVATE OR PERISH

I never think of the future. It comes soon enough.

ALBERT EINSTEIN (1879–1955)

But, be it for better or for worse, if ever the old industries are changed, so will the old population be changed. The new types may be sturdy, heavy-limbed farmers, healthy fruit-pickers and fruit packers; or they may be factory folk. But, good or bad, they will not be the old types. If ever the wool industry of Australia were to be not merely supplemented, but supplanted, by other industries, there would go with it some of the most vital types which, a few years after our visits, went to make up the men of Anzac.

CEW BEAN, PREFACE TO ON THE WOOL TRACK, 1962

The growing of wool created Australia's economic wealth. The human qualities and skills required to grow it helped form the unique Australian character. If it had not been for sheep and the men who followed them, much of arid inland Australia may never have been settled. Yet at the start of the 21st century the future for woolgrowers in Australia remained uncertain.

There was a collective sigh of relief around the bush when, on Thursday August 9, 2001, the last bale of wool from the stockpile was sold. The wool mountain accumulated during 1989 and 1990 under the Reserve Price Scheme had stood at 4.7 million bales in January 1991, with an attached debt of $2.8 billion. Its malign influence had depressed wool prices for most of the 1990s, and locked many unfortunate growers into grinding poverty. Their reaction to all those lean years had been to slash sheep numbers and turn to doing other things. So in 2004 the national sheep flock was at its lowest level since the years immediately following the Second World War, and was still shrinking.

In response to reduced supplies wool prices jumped, and during 2002 the market indicator lifted above 1000 c/kg clean for the first time in more than 14 years. Western Riverina clips were averaging more than $1500 per bale, compared to around $500 during the grim years. Growing wool was profitable again. In the past, such a quick turnaround in prices had triggered a return to wool growing and increases in the national flock. It had done so in the 1970s, but this time something entirely different was happening.

In the winter and spring of 2001 sheep prices boomed. At the July sheep sale in Hay, New South Wales, the offering of 23,000 was one of the smallest in recent memory, and young ewes sold to $64.50 and wethers to $59. The average of $42 a head for wethers on the day was $17 a head in the previous year. As a sign of good times returning, the top-priced ewes at $64.50 – sold by the Housten family of *Budgewah* – were the progeny of old ewes bought the previous year for $7.50 a head. At a sale in Jerilderie the same week one delighted vendor – Ian Seidel from *Buckinbong* at Narrandera – announced he was going to frame the sale invoice after his pen of young ewes made $67, double his price of the previous year. It was a powerful symbol of the times that the same Jerilderie sale witnessed the complete dispersal of the well-known *Tarabah* flock, which had been carefully bred up by three generations of the Coughlan family near the village of Morundah. The current owner was switching to cattle and cropping. Away to the north in western Queensland five famous stations that had run renowned Merino flocks since the 19th century were switching over to cattle – *Malvern Hills, Athol, Northampton Downs, Terrick Terrick* and *Eniskillen*.

The slump of the 1990s was just too protracted and cut too deep, the restructuring too chaotic; the robust confidence that woolgrowers had once felt about their future had evaporated, and they were taking advantage of booming world demand for sheep meat to quit their flocks – probably forever. It seems unlikely that the Australian sheep flock will ever return to a figure around 150 million. Property owners have lost interest in sheep, for reasons more complex than the simple commercial aspects of them being labour intensive to run. The husbandry of sheep does not conform to the modern need for speed of action, measurable short-term results, predictability of returns, and tight control of activities. Those in their 20s and 30s who had grown up during the 1990s have learned to dislike sheep. For them the control of powerful and sophisticated farm machinery is more attractive and satisfying, particularly as it involves use of technology such as computers and satellites.

In the early 1960s I visited the Narrabri district to class sheep for a mixed farmer named Mervyn Quick, and I was not looking forward to it – very often sheep on a wheat farm played second fiddle to cropping and were not well cared for. My heart sank when I realised there was only the elderly owner, a broken-down kelpie sheep dog and myself to put 3000 sheep through a tumbledown set of sheep yards in the day available. Then I spotted the sheep – they were magnificent. In full wool, they moved with the slow rolling gait of animals carrying eight-kilogram fleeces. The owner tied his old dog to the back fence and opened all the internal gates; then the sheep just kept moving forward in a steady, unfussed stream, never jibbing or turning back. So quiet, well fed and content were they that as soon as they passed out of the classing race they just lay down and went to sleep. It turned out to be a delightful day handling fine stock as they should be handled – quietly and without fuss. Mervyn Quick was a superb stockman. By 2002 there were few people prepared to invest the years it took him to learn the gentle craft of stockmanship, to understand how to handle sheep, to anticipate their needs, and to feed them properly. In a fast-moving world sheep had become anachronisms to many young people on the land.

Industry rationalisation continues. In most wool-growing regions of Australia there is the melancholy sight of abandoned homesteads and derelict woolsheds gradually falling apart or being overwhelmed by encroaching scrub. There were 206,000 agricultural operations in Australia in 1955, by 1989 this had shrunk to 126,000, and is still falling. A generation was moving out; many sons and daughters of soldier settlers, baby boomers in their 50s, have had enough. They have been well-educated, have taken city or town jobs, and are not interested in returning to the family property and continuing their parents' battles. Properties are being amalgamated. In the north-west corner of New South Wales in the area serviced by Broken Hill, 35 former owner-occupied properties were controlled by just 11 families at the turn of the 20th century. The owner and his wife – with the help of contract labour at stress times – were running these larger properties, which once would have employed 30 or more permanent hands. This has nothing to do with globalisation, competition policy or trade deregulation, but is because of the use of technology – helicopters, light aircraft, motorcycles, radios and computers. By 2001 more than 76 per cent of woolgrowers were using computers, one-third were using email and 92 per cent had facsimile machines. But rationalisation has not gone far enough to give woolgrowers an efficient marketing system, and they are still too reliant on a volatile spot auction market to sell their clips. In other commodity-based industries the scale of production units has more or less kept pace with the scale of industrial user units, so the respective strengths of buyers and sellers are in better balance. In that situation the parties are negotiating direct contracts, often on a forward pricing basis, and little production is left at the mercy of a volatile spot market.

During the years covered in this book woolgrowers were generally ill-served by their political representatives at all levels. The changes and rationalisation being wrought at the start of the 21st century seemed likely to alter the rural political landscape forever. There is increasing polarisation between woolgrowers. At one end a rising number of professional growers are adopting a business approach; at the other a decreasing number are running sheep as a by-product or a hobby and supplementing their incomes with off-farm work. This sector is often willing to accept lower returns and a diminished lifestyle that would not be acceptable to other business owners. The 'average' farm no longer exists, and this situation is testing the relevance of the traditional grower organisations. As *The Land* columnist Peter Austin commented in April 2001: 'Leading farmers are likely to be well-educated, professionally advised, large in scale, well-connected in terms of marketing their produce, entrepreneurial by nature, and used to making their own way in the world. As such, they will see the world differently, and have very different needs, to someone on your average farm battling to make ends meet or to keep up with technological change.' Larger growers are starting to question the value of belonging to state or federal farmer organisations, just as trade unionists are questioning the relevance of their old structures. It seems as if many are likely to go their own way, leaving the old bodies with a constituency consisting of a 'rump' of marginal producers requiring an ever-more radical agenda.

The biggest question that hangs over the heads of growers is whether, with reduced world supplies, wool will be able to remain relevant in the world textile markets. Wool holds only a small share of the apparel textile

AWI's MerinoFresh technology allows suits to be washed under a shower then drip-dried without creases ready to wear again ***(AWI)*****.**

market, and is under constant pressure from synthetics, which every year are getting closer to replicating the core qualities of wool – drape, texture, softness, insulation and longevity. With supplies shrinking further, the investment decisions of processors faced with a wide choice of fibres will be critical.

Clothes made from wool will always be more expensive than those made from competing fibres because of the long supply chain, expensive processing and volatile raw material prices. Customers for clothing, as for other consumer products, have a bewildering number of choices. The internet can bring nearly the full range of offerings to a computer screen at a click, almost anywhere in the world; clothing, once an emotional purchase, is increasingly becoming a commodity that customers shop for strictly on the basis of price. So consumers need to be convinced they are getting additional value when they buy woollen clothes. Wool needs a trusted brand identity to differentiate itself and avoid burial in the ruck. The decision to abandon grower-funded generic promotion of the fibre that was made following the

McLachlan taskforce inquiry may have been a grave error. The trusted Woolmark symbol, recognised by consumers around the world, risks losing its power through lack of nourishment from advertising.

Just as wool was retreating from generic promotion, other fibres were increasing their efforts. Cotton remained the giant of the textile fibre market; in 2000 its share was running at around 40 per cent. In November 2000 the International Cotton Advisory Committee held a meeting in Cairns, where delegates heard from the managing director of the giant agribusiness financier Rabobank, Adrian De Young. He warned: ' … those who believe cotton will always maintain market share as a natural fibre are dreaming. The answer is marketing, marketing, marketing.' He gave as an example how Cotton Incorporated in the US, founded in 1970 as a grower-funded organisation, had lifted cotton's share of the domestic market from 34 per cent in 1975 to 61 per cent. Cotton Incorporated senior director Mark Messura could have been reading from an International Wool Secretariat operating manual from the 1980s when he explained how they did it. He said their success could be attributed to a 'push/pull' marketing strategy, focused on 'pushing' cotton production and textile innovations into the market, and 'pulling' back added consumer demand through advertising and promotion. He said a key part of the strategy had been development of the 'Cotton Seal' symbol used to identify and brand products at retail level. Another speaker – Andrew Macdonald, president of the Brazilian Cotton Association – called for an international 'Cotton Seal': 'We need a mark for us all to rally to, and to be the initial focus of generic promotion.' Woolgrowers live in hope that the decision to abandon generic promotion will not lose them their own precious Woolmark symbol and the investment that had been put behind it since 1964.

Wool growing, which helped to build the Australian nation, faces an uncertain future at the start of the 21st century. The 1970s eroded much of the veneration city folk had once felt for the woolgrower; the behaviour of growers during the 1990s probably demolished the last vestiges of it. Short of a massive restructuring of Australia's rural industries, it looks unlikely that wool will ever recover its major role in the national economy. Yet there remains something in Australian culture that likes to think that outback sheep stations – and the worthy, hard-working people who run them – are still out there. And they are. The gloomy prediction that CEW Bean made in 1962 about the loss of national character should the wool industry be supplanted has not come to pass. Though the industry he knew in those and earlier years has been partly dismantled, the human qualities that made it great still exist. Resolute and tenacious, pastoralists remain indomitable in the face of difficulties that would defeat lesser people. Persistent and determined, they have endured, sometimes to their own detriment, to survive despite all that markets, seasons and the mismanagement of politicians have thrown at them. They helped to make Australia.

EPILOGUE

THE HUB IN THE WHEEL

by Amanda Burdon

The real solution rests with on-farm efficiency. This efficiency is not restricted to the size of the flock. Small flocks can be just as efficient as large flocks.

ALEC MORRISON (1935–2004)

The industry that Alec Morrison knew and loved has continued to evolve since his death in 2004 and evolve it must if wool is to have a future. As the decade draws to a close, the size of the Australian flock has plummeted to a 90-year low – of approximately 79 million sheep – and wool profits have been shorn. Trends that Alec identified five years ago regarding pricing volatility, wool's declining competitiveness, the rise of more profitable land uses and increasing diversification have continued to impact heavily on wool production and returns. Soaring input costs and prolonged drought conditions across vast swathes of the continent have afforded many cash-strapped growers little option but to develop alternative income streams, and to opt for dual-purpose sheep breeds. When coupled with the divisive debate over mulesing, increased environmental accountability and other global market forces, wool-growing in Australia, especially on the smaller scale, has become an increasingly challenging proposition.

'The parameters of food and fibre production are shifting because of the demand for biofuels; diesel prices; climate variability; the credit crunch and the lack of finance; record land, herbicide and fertiliser prices;

and increased animal welfare and environmental pressures,' said fourth-generation woolgrower and agricultural consultant Peter Schuster, of Dubbo, New South Wales. 'This has all conspired to create an atmosphere of great uncertainty in the wool industry.'

Yet, despite the seemingly grim outlook, there are still plenty of dyed-in-the-wool producers prepared to meet the challenges head-on; people with an enduring sense of tradition and optimism who continue to demonstrate the kind of determination and innovation that the industry was founded upon – and which Alec epitomised. The national flock may never fully recover from the onslaughts of the past decade, but there is little question that Australia continues to produce some of the most desirable and luxurious fibres in the world. From Bourke to the pastoral areas, the Gippsland coast to the Monaro, the superbly adapted Merino remains the backbone of our nation's sheep industry and a vital part of our heritage.

That's not to say that the composition of the Australian flock and business models for growing wool will not continue to change. Indeed, diversification has been at the heart of the survival strategies employed by many growers in recent times. As overall wool profits have declined in the face of sustained meat and grain prices, there has been a pronounced shift towards sheep-meat production, broad-acre cropping and beef production. By 2007-08, lamb production had jumped by almost 50 per cent, and more than 60 per cent of properties that once grew wool exclusively were mixed farming enterprises. 'The fellows who are succeeding at wool-growing have a self-replacing Merino flock, terminal sires for meat production, grow some cereals and have also been active on the share market,' said leading stud classer and consultant Michael Elmes, who managed *Wanganella* while Alec was at the helm of F.S.Falkiner.

A recently elected Australian Wool Innovation board member and upholder of the proud family name and traditions, George Falkiner believes that intense competition for land use is largely responsible for driving sheep numbers down. 'It all comes back to profitability per hectare or lifestyle choices,' he said. 'Labour-intensive sheep farms are not favoured by the younger generation and they don't lend themselves so readily to automation, as in other agricultural sectors. Wool remains a unique product, but the meat side now underpins many wool operations, giving growers the confidence they need to continue producing wool. But while there's money to be made in meat, behind every meat breed you will find a Merino ewe.'

For, despite the deterrents, Merino wool remains an attractive package. 'It's still the blue-chip share,' said Cam Munro, general manager of the influential stud *Egelabra*, near Trangie, New South Wales. 'A lot of people have tried exotics in the western region, but there is still a big percentage of pure Australian sheep. It is not a high-risk investment like cropping. It can be the hub in the wheel of many commercial operations. It's all about spreading your risk, especially during dry years, when the Merino has shown its adaptability and good lambing percentages. Wool is one of the few commodities that will yield a return in drought.'

Those growers that have held on to their Merino-breeding ewes – and they continue to dominate the Australian flock – have become much more speculative in their joinings. Some are alternating between wool and meat production through sire breed selection while others have moved toward selecting Merinos that demonstrate dual-purpose characteristics – large, well-muscled animals producing fine apparel wool. Cam Munro believes that this capacity demonstrates the true versatility of the Merino breed – that its strong genetic foundation enables breeders to respond swiftly to market signals without jeopardising long-term wool production. But a growing number of sheep breeders are turning to plainer, easier-care breeds with better fertility that are capable of producing early-maturing lambs. The growth in dual-purpose breeds has, naturally, had a major bearing on wool quality, producing a greater proportion of coarser wools (greater than 26.5 microns) suitable for carpets and upholstery. According to the Australian Wool Testing Authority (AWTA), between 1992–93 and 2007–08 that proportion of the clip (primarily from non-Merino sheep) increased from 6.5 per cent to 10.6 per cent. The decline in wool volume nationally from July 2008 to March 2009 reflects this change in flock structure, towards more cross-breeding and broader wool production.

But at the opposite end of the scale, the quest to produce finer, heavier, and potentially more valuable fleeces is largely undiminished. Over the same period, the AWTA reports that the proportion of superfine wool (less than 19.6 microns) increased more than fourfold, from 7.6 per cent to 35 per cent of the national clip. And it's not just the fine and superfine producers who are lifting their game. Wether trials show that even medium-wool producers have taken, on average, two microns off their clips, while maintaining fleece weights, and that wool in the intermediate categories fell, as a proportion of the clip, from 86 per cent to 56.2 per cent over the same period.

However, it's not just fibre diameter, staple strength and fertility imperatives that are guiding breeding strategies. Today's growers increasingly select for easier-care animals with better resistance to disease and parasites, and major advances in sheep genetics have helped to develop bare-breeched sheep that are less costly to run. Laurie Piper contends that 15 years ago it's likely Alec would have culled many of these sheep, but believes the tools that have enabled such breeding developments – like the national sheep genetic evaluation scheme – are invaluable to contemporary breeders. 'They can now use objective information to breed for a vast array of traits, not just for fibre diameter and fleece weight, but also for other important characteristics like staple strength, style, handle, colour and parasite resistance,' Laurie said. 'There are still people who breed by hand and eye assessments, but others are

Some growers are selecting large-framed, heavy-cutting, plainer-bodied Merinos that demonstrate dual-purpose meat/wool characteristics *(Schuster Images)*.

using sheep genetics, estimated breeding values, pedigree information, sire evaluation and wether trial results more and more. The availability of this information is changing the face of sheep breeding.'

The broader commercial and consumer imperatives influencing breeding are most apparent when it comes to developing flystrike resistance. The international furore over the practice of mulesing, in the wake of the public campaign waged by animal welfare group People for the Ethical Treatment of Animals (PETA), has cast the spotlight on sheep husbandry like never before and tested reputations and relations. In urging the world's largest retailers to boycott Australian mulesed wool, PETA has ignited a debate over what constitutes ethical wool production. The Australian wool industry has responded by committing to phase out mulesing by December 2010 (although most now admit this deadline cannot be met) and pledging to investigate a range of alternative flystrike management strategies, including breech and tail clips, the use of analgesia, selective breeding, new manual care and husbandry techniques, and chemical fly-control measures.

But the mulesing debate has fractured an already splintered industry. Some growers protest that the ban will sound the death knell of the industry; still others welcome what they claim are long overdue improvements to sheep management and genetic selection. The crisis has certainly renewed interest in breeding low-wrinkle and bare-breeched sheep, which people like former CSIRO scientist Jim Watts have spent years refining. 'We all want an alternative to mulesing but it has to be an economical one otherwise we will see a lot of people leave the industry especially in the pastoral belt,' said Cam Munro. 'It's not as simple as using a plainer-bodied sheep to prevent breach strike. All tests on non-mulsed plain-bodied sheep have been carried out in the worst drought in history. They are yet to be tested in a five- to six-year run of wet years like the 90s. The plain-bodied more open sheep may be more susceptible to body strike in this time. The inside country may be able to get away without mulesing through double crutching and extra chemical use, but we can't get away from mulesing at this stage in the pastoral region, where growers don't see their sheep for months on end and have no staff.'

Science has not yet provided all the solutions to the challenges the industry faces and nor has it supplanted the tradition that Alec sought to celebrate. As he predicted, those studs that have successfully '... integrated traditional craft skills with objective measurement in a commonsense way …' are the ones that have claimed market share, maintaining some of the 'mystery and magic' that he so admired. George Falkiner believes that some of today's best stud classers and masters are 'every bit as knowledgeable as the old-time greats' and skilfully combine tradition and modern technology. 'The new generation of stud masters are performance breeders, who rely on measurements, but also the magic of touch and feel,' he said. 'Nobody has bred an Australian grand champion on Australian Sheep Breeding Values yet. The eye for feet, backlines, horn settings and the like – we are yet to put a number on that.'

The market is also yet to quantify what it's prepared to pay for wool produced in an 'ethical' or environmentally-friendly fashion. The Australian Wool Exchange has introduced new categories for non-mulesed wool and growers can voluntarily declare their wool as such, but projected premiums are yet to be realised. However, some anticipate that welfare standards may become a point of difference and a major determinant of future sales. To satisfy consumer demand, the Australian Farm Institute has even suggested that, if not mandatory standards, then at least voluntary accreditation and associated labelling may be demanded of Australian livestock industries like the wool industry in future.

International signals suggest that wool's natural appeal is no longer enough to secure its market success. Environmental management and performance along the convoluted supply chain is being subjected to increased scrutiny, according to global natural resource management principles. The condition of the soils on which wool is grown, the health of pastures and native vegetation, and river and water quality, are all tenets of an emerging philosophical and business model known as ecosystem accounting. That Japanese apparel giant Onward Kashiyama has promised a percentage of its Merino product sales to help landholders to improve their environmental, social and economic performance through its support of a formal, certificate-based environmental program is testimony to this shift in thinking. Similar projects and business partnerships recognise that sustainable farming and animal welfare practices are increasingly important influences on retail spending.

Of course, this may all augur well for an essentially natural, renewable and biodegradable product, provided production is demonstrated to be sustainable. Woolgrowers who use little or no chemicals, do not mules, and who practise holistic resource management have already positioned themselves to capture markets for accredited organic wool brands. Others are also seeking to capitalise on the opportunities borne of the future carbon economy. 'For woolgrowers, as in any business, we talk increasingly of needing to have a social licence to operate,' said Adrian Kloeden, the former CEO of AWRAP under Alec's chairmanship. "Anyone who can respond to that has a market advantage. Along the wool supply chain, there may be opportunities to make individual pipelines for the certification of certain types of wool, grown in a certain way. Growers will need to better understand marketing, branding and global market demands."

But as the Australian Bureau of Agricultural and Resource Economics (ABARE) recently observed: '... the continued success of the wool industry depends not only on the relative price competitiveness of wool but also on its ability to remain in demand as a desirable fibre in products sold to consumers'. With wool commanding just 2.5 per cent of the world's total textile market (down from about six per cent in the early 1990s) or about eight per cent of the apparel fibre market (down from 14 per cent in the early 1990s) there is much ground to be recovered in understanding and meeting market needs and expectations.

Since Australian Wool Services and AWI merged in 2007, AWI has devoted considerable resources to improving industry-wide cooperation, research and development, and invested heavily in product development and wool marketing. AWI has pledged to spend more than $120 million from 2008–2011 re-launching Australian wool under a new premium Merino brand, aimed at positioning it internationally '... as the quality fibre of choice for apparel products'. It is a bold and ambitious effort to appeal to the fast-growing global sports and active-wear, as well as ladies fashion segments. There are few who would argue that effective marketing is fundamental to wool's future and these commitments, according to Laurie Piper, are some cause for optimism. 'We are finally seeing a lot of blending of wool, to produce next-to-skin fabrics, and rebranding wool as Merino was a clever stroke that allows the industry to talk about wool as if it were a new fibre altogether,' he said. 'For the first time, the industry is truly catering for what consumers want and maybe even influencing what they want by creating new and interesting products.'

Today's wool industry is vastly different to the one that Alec toiled in as a young man and, indeed, different again from the one he presided over. The golden era of wool may have passed but, 200 years on, this uniquely Australian industry survives – and passionate, progressive, adaptable growers along with it. George Falkiner recalls Alec once asking him whether he thought the great sheep studs would ever return. 'I replied that I thought the smaller, family-operated studs would become more dominant, because they are the most adaptive, and that has proven to be true,' George said.

The living fibre that can be soft and strong, warm yet cool, water-repellent as well as absorbent, has often been described as highly responsive. At its divine best, wool shines. The same could be said for those who grow it. As the national flock size continues to fall and global forces extinguish some of the international demand for wool, the ingenuity and innovation for which the industry is renowned will once again need to come to the fore. 'Amorphous change has always characterised the industry and change will continue; it's a cyclical industry that never stays the same as we battle nature and try to convert rain into money,' said Barry Walker, a board member when Alec chaired AWRAP. 'We just need to accept change, grab the opportunities and maximise the outcomes.' Growers are certainly being forced to develop new skills as they move away from a commodity culture and implement new business models, adapt sheep breeding to meet current consumer demands, and work to reinvigorate the international demand for wool with exciting new fabrics and designs. Across the landscape of wool production, the process of evolution continues.

BIBLIOGRAPHY

AI Alletzhauser, *The House of Nomura,* Bloomsbury, London, 1990.

HB Austin, *The Merino – Past, Present and Probable,* Graham Book Company, Sydney, 1943.

Allen Barnard (Ed), *The Simple Fleece – Studies in the Australian Wool Industry,* ANU/MUP, Melbourne, 1962.

CEW Bean, *On the Wool Track,* Sirius Books, 1963 Edition. Originally published by Angus and Robertson, Sydney, 1910.

Brian Coman, *Tooth and Nail – The Story of the Rabbit in Australia,* Text Publishing, Melbourne, 1999.

Jill Ker Conway, *The Road from Coorain,* Mandarin Paperbacks, London, 1990.

Barbara M Duncan, *My Chequered Childhood,* John M Duncan, Bayswater (circa 1980).

Suzanne Falkiner, *Haddon Rig – The First Hundred Years,* Valadon Publishing, Sydney, 1981.

Cuthbert Fetherstonhaugh, *After Many Days,* E.W.Cole, Melbourne, 1917.

Peter Forrest, *A Rush for Grass,* Murranji Press, Darwin,1988.

John Galsworthy, *Maid in Waiting,* , William Heinemann, London, 1931.

John Ryrie Graham, *A Treatise on the Australian Merino,* Clarson, Massina and Co., Melbourne, 1870.

Robert Graves and Allen Hodge, *The Long Weekend,* Penguin, UK, 1971.

Tim Hewat, *Golden Fleeces – The Falkiners of Boonoke,* Bay Books, Sydney, 1981 and *Golden Fleeces II – The Murdoch Years at Boonoke,* Bay Books, Sydney, 1988.

F Eric Hitchins, *Tangled Skeins, A History of Australian Wool Marketing,* Robertson and Mullins, Melbourne, 1956 and *Skeins Still Tangled – Australian Wool Marketing 1952–1972,* Hawthorn Press, Melbourne, 1972.

Rowand Jameson, *Under the Chloroform Tree: from jackaroo to studmaster,* Shoestring Press, Wangaratta, 1984.

Kathielyn Job, (editor), *Merino Musters,* published in Dubbo by the author, 1985.

FF Knight, *A Memoir,* Chevron Media, Double Bay, 1980.

B Macsmith, *Quench Not the Spirit,* Hawthorn Press, Melbourne, 1972.

Charles Massy, *The Australian Merino,* Viking O'Neill, Penguin Books Australia, Ringwood, 1990.

Fred Moylan, *U-Turn for Wool,* New Street Books, Melbourne, 1993.

H Munz, *The Australian Wool Industry,* Cheshire Publishing, Melbourne, 1968.

Nyngan Historical Society, *Nyngan on the Bogan,* published by the Society, 1986.

Joan Austin Palmer, *Memories of a Riverina Childhood,* New South Wales University Press, Kensington, 1993.

EH Pearse, *Sheep and Property Management,* (seventh edition), 1951, The Pastoral Review Pty Ltd, Melbourne.

Murray T Pheils, *The Return to Coree – The Rise and Fall of a Pastoral Dynasty,* Allen and Unwin, St Leonards, 1998.

Charles W Russell, *Country Crisis,* W.R.Smith and Paterson, Brisbane, 1976.

Douglas Stewart and Nancy Keesing (editors), *Old Bush Songs,* Angus and Robertson, Sydney, 1970.

Alec Watson, *A Stock Agent Remembers,* George Alexander Watson, Euroa, 1983.

Denis Winter, *The First of the Few: Fighter Pilots of the First World War*, Allen Lane, London, 1983.

NEWSPAPERS AND PERIODICALS

Dalgety's Review, files held at *Boonoke*, Deniliquin, New South Wales.

The Land, files held at Public Library, Sydney.

The New South Wales Pastoral Directory, files held at *Boonoke*, Deniliquin, New South Wales.

The Pastoral Review, files held at *Boonoke*, Deniliquin, New South Wales.

The Stock and Land, files held at La Trobe Library, Melbourne.

The Weekly Times, files held at La Trobe Library, Melbourne.

The Woolmakers, (editor Tim Hewat), magazine editions from 1979–99 private collection.

OTHER REFERENCES

Australian Stud Merino Flock Register, various editions.

Bicentennial Wool Conference, *Conference Papers,* Sydney, 1988.

Centre For International Economics Report, *Wool into the 21st Century*, Canberra, 1990.

First World Merino Conference, *Conference Papers*, Melbourne, 1982.

McLachlan Wool Taskforce Report, *Diversity and Innovation for Australian Wool*, 1999.

Professional Sheepclassers' Association of Australia, *Proceedings of 1965 and 1969 Conventions*.

Sir Richard Randall and Committee Report, *The Australian Wool Industry 1971–72*.

Dr Bob Richardson, *The Politics and Economics of Wool Marketing, 1950–2001* in *The Australian Journal of Agricultural and Resource Economics*, March 2001, Australian Agricultural and Resource Economics Inc. and Blackwell Publishers Ltd, Oxford, UK.

Sir William Vines and Committee Report, *The Australian Wool Industry – Recommendations for the Future*, March 1991.

OB Williams, *The Riverina and its Pastoral Industry 1860–69*, CSIRO pamphlet, 1960s.

Barratta archives (held at *Barratta*, Deniliquin New South Wales): books of account, wages books, managers' reports, correspondence to head office, ram sales books, general correspondence, store books, diaries, telegram Books.

Falkiner archives (held at *Boonoke*, Deniliquin, New South Wales): books of account, letter books, annual reports, board minutes, paddock books, diaries, general correspondence, stud sheep records, ram sales books, ram sales correspondence, telegraph books, wages books, wool books.

Otway Falkiner's personal correspondence.

Letter books held at Boonoke, Deniliquin, New South Wales.

Buckiinguy Station Annual Report, 1931.

The Australian Pastoral Company – internal memos and correspondence.

Alec Morrison's personal correspondence, diaries and reports.

JB Dowling's diaries and personal papers.

Jim Young's unpublished manuscript, 1994.

NOTES

CHAPTER ONE: JACKAROOS

1 Peter Forrest, *A Rush for Grass*
2 Rowand Jameson, *Under the Chloroform Tree*
3 *The Pastoral Review*
4 *The Pastoral Review*
5 *A Rush for Grass*
6 *Barratta* archives
7 Kathielyn Job (editor), *Merino Musters*
8 *A Rush for Grass*

CHAPTER TWO: FAR-FLUNG FIEFDOMS

1 Robert Graves and Allen Hodge, *The Long Weekend*
2 *The Pastoral Review*
3 Joan Austin Palmer, *Memories of a Riverina Childhood*
4 Jill Ker Conway, *The Road to Coorain*
5 *The Pastoral Review*
6 *The Pastoral Review*
7 Rowand Jameson, *Under the Chloroform Tree*
8 *The Long Weekend*
9 *Under the Chloroform Tree*
10 *The Long Weekend*
11 Murray Pheils, *Return to Coree*
12 *The Long Weekend*

CHAPTER THREE: THE THREADBARE THIRTIES

1 Alec Watson, *A Stock Agent Remembers*
2 Peter Forrest, *A Rush for Grass*
3 Rowand Jameson, *Under the Chloroform Tree*
4 *A Rush for Grass*
5 CEW Bean, *On the Wool Track*
6 Sir William Vines, from a speech made in 1988
7 Murray Phiels, *Return to Coree*
8 Falkiner archives
9 Letter to *The Pastoral Review*

10 *Under the Chloroform Tree*
11 *The Pastoral Review*
12 *The Queensland Graziers Review*
13 *A Rush for Grass*
14 *A Rush for Grass*

CHAPTER FOUR: WARTIME IN THE HOMESTEADS

1 Peter Forrest, *A Rush for Grass*
2 *The Pastoral Review*
3 *The Pastoral Review*
4 *The Pastoral Review*

CHAPTER FIVE: THE FABULOUS FIFTIES

1 CW Russell, *Country Crisis*
2 Rowand Jameson, *Under the Chloroform Tree*
3 Suzanne Falkiner, *Haddon Rig – The First Hundred Years*
4 *Under the Chloroform Tree*
5 Murray Pheils, *Return to Coree*

CHAPTER SIX: A DECADE OF CHANGE

1 *The Weekly Times*
2 Peter Forrest, *A Rush for Grass*
3 *A Rush for Grass*
4 Fred Knight, *Memoirs*
5 *Memoirs*

CHAPTER SEVEN: CATTLE BOOM ... AND BUST

1 *The Weekly Times*

CHAPTER EIGHT: STATION TUCKER

1 Murray Pheils, *Return to Coree*
2 Henry G Lamond, Letter to *The Pastoral Review*
3 Joan Austin Palmer, *Memories of a Riverina Childhood*

CHAPTER NINE: FUELLING CHANGE

1 Fred Knight, *Memoirs*
2 *The Pastoral Review*
3 Rowand Jameson, *Under the Chloroform Tree*
4 Denis Winter, *The First of the Few*
5 *Barratta* archives
6 Joan Austin Palmer, *Memories of a Riverina Childhood*
7 *Under the Chloroform Tree*
8 Letters to *The Pastoral Review*
9 Letter to *The Pastoral Review*

10 Peter Forrest, *A Rush for Grass*
11 Letter to *The Pastoral Review*
12 Letter to *The Pastoral Review*

CHAPTER TEN: COMPANY MEN

1 Allen Barnard, *The Simple Fleece*
2 *The Simple Fleece*
3 Rowand Jameson, *Under the Chloroform Tree*
4 Peter Forrest, *A Rush for Grass*
5 *A Rush for Grass*
6 John Ryrie Graham, *A Treatise on the Australian Merino*
7 *Under the Chloroform Tree*
8 *A Rush for Grass*
9 *A Rush for Grass*
10 *The Pastoral Review*
11 *Under the Chloroform Tree*
12 Barratta archives
13 *Under the Chloroform Tree*
14 Falkiner archives
15 *A Rush for Grass*
16 *A Rush for Grass*

CHAPTER ELEVEN: SOLDIER SETTLERS

1 *The Pastoral Review*
2 *The Weekly Times*
3 Peter Forrest, *A Rush for Grass*
4 *A Rush for Grass*
5 Jill Ker Conway, *The Road from Coorain*
6 *The Road from Coorain*
7 *The Road from Coorain*
8 A paper prepared by the Bank of New South Wales

CHAPTER TWELVE: RABBITS, RABBITS, RABBITS

1 Brian Coman, *Tooth and Nail*
2 *Tooth and Nail*
3 *Tooth and Nail*
4 *Tooth and Nail*

CHAPTER THIRTEEN: DROUGHT TAKES HOLD

1 *The Pastoral Review*
2 *The Pastoral Review*
3 *The Pastoral Review*
4 *The Pastoral Review*
5 Falkiner archives

CHAPTER FOURTEEN: THE GREAT DYNASTIES CRUMBLE

1 Suzanne Falkiner, *Haddon Rig – The First Hundred Years*
2 Fred Knight, *Memoirs*
3 *Memoirs*
4 John Galsworthy, *Maid in Waiting*

CHAPTER FIFTEEN: STUD BUSINESS

1 John Ryrie Graham, *A Treatise on the Australian Merino*
2 *The Pastoral Review*
3 HB Austin, *The Merino – Past, Present and Probable*
4 *A Treatise on the Australian Merino*
5 *The Merino, Past, Present and Probable*
6 Cuthbert Featherstonhaugh, *After Many Days*

CHAPTER SIXTEEN: WOOLLY-MINDED FADS

1 Letters to *The Pastoral Review*
2 HB Austin, *The Merino – Past, Present and Probable*

CHAPTER SEVENTEEN: THE 'BIG SIX'

1 Peter Forrest, *A Rush for Grass*
2 Joan Austin Palmer, *Memories of a Riverina Childhood*
3 Kathielyn Job (editor), *Merino Musters*
4 Suzanne Falkiner, *Haddon Rig – The First Hundred Years*
5 Falkiner archives
6 Fred Knight, *Memoirs*
7 Barbara M Duncan, *My Chequered Childhood*

CHAPTER EIGHTEEN: HEYDAY AT THE FAIR

1 *The Pastoral Review*
2 Letter in the possession of the author
3 Kathielyn Job (editor), *Merino Musters*

CHAPTER TWENTY: SCRUTINY OF SCIENCE

1 Proceedings of the Bicentennial Sheep and Wool Conference 1988
2 *The Pastoral Review*
3 *The Pastoral Review*

CHAPTER TWENTY-ONE: A CLASS APART

1 John Ryrie Graham, *A Treatise on the Australian Merino*
2 Tim Hewat, *Golden Fleeces*
3 HB Austin, *The Merino – Past, Present and Probable*
4 Charles Massy, *The Australian Merino*
5 Falkiner archives
6 *The Pastoral Review*

7 Fred Knight, *Memoirs*
8 Australian Sheep Classers Convention, *Proceedings*

CHAPTER TWENTY-THREE: THE GRAZIER/FARMER FAULTLINE

1 F Eric Hitchins, *Tangled Skeins*

CHAPTER TWENTY-FOUR: SIR WILLIAM GUNN

1 F Eric Hitchins, *Skeins Still Tangled*
2 *Skeins Still Tangled*
3 *Skeins Still Tangled*
4 CW Russell, *Country Crisis*

CHAPTER TWENTY-FIVE: CAST OFF THE SHEEP'S BACK

1 Letter to *The Land*

INDEX